THE LIBRARY
ST. MARY'S COLLEGE OF MARYLAND
ST. MARY'S CITY, MARYLAND 20686

T5-BCA-518

GENDER, SCIENCE AND MATHEMATICS

Science & Technology Education Library

VOLUME 2

SERIES EDITOR

Ken Tobin, *Florida State University, Tallahassee, Florida, USA*

EDITORIAL BOARD

Beverly Bell, *University of Waikato, Hamilton, New Zealand*

Reinders Duit, *University of Kiel, Germany*

Kathlene Fisher, *San Diego State University, California, USA*

Barry Fraser, *Curtin University of Technology, Perth, Australia*

Chao-Ti Hsiung, *National Taipei Teachers College, Taipei, Taiwan*

Doris Jorde, *University of Oslo, Norway*

Michael Khan, *Centre for Education Policy Development, Braamfontein, South Africa*

Vince Lunetta, *Pennsylvania State University, University Park, Pennsylvania, USA*

Pinchas Tamir, *Hebrew University, Jerusalem, Israel*

SCOPE

The book series *Science & Technology Education Library* provides a publication forum for scholarship in science education. It aims to publish innovative books which are at the forefront of the field. Monographs as well as collections of papers will be published.

The titles published in this series are listed at the end of this volume.

Gender, Science and Mathematics

Shortening the Shadow

Edited by

LESLEY H. PARKER, LÉONIE J. RENNIE and BARRY J. FRASER

Curtin University of Technology, Perth, Australia

KLUWER ACADEMIC PUBLISHERS
Dordrecht / Boston / London

Library of Congress Cataloguing-in-Publication Data

Gender, science and mathematics : shortening the shadow / edited by
  Lesley H. Parker and Leonie J. Rennie and Barry J. Fraser.
      p.   cm. -- (Science & technology education library ; v. 2)
  Includes bibliographical references.
  ISBN 0-7923-3535-X (alk. paper)
      1. Women in science.   2. Women in mathematics.   3. Educational
  equalization.   I. Parker, Lesley H.   II. Rennie, Leonie J.
  III. Fraser, Barry J.   IV. Series.
  Q130.G46 1995
  507--dc20   95-30191

ISBN 0-7923-3535-X

Published by Kluwer Academic Publishers,
P.O. Box 17, 3300 AA Dordrecht, The Netherlands.

Kluwer Academic Publishers incorporates
the publishing programmes of
D. Reidel, Martinus Nijhoff, Dr W. Junk and MTP Press.

Sold and distributed in the U.S.A. and Canada
by Kluwer Academic Publishers,
101 Philip Drive, Norwell, MA 02061, U.S.A.

In all other countries, sold and distributed
by Kluwer Academic Publishers Group,
P.O. Box 322, 3300 AH Dordrecht, The Netherlands.

Printed on acid-free paper

All Rights Reserved
© 1996 Kluwer Academic Publishers
No part of the material protected by this copyright notice may be reproduced or
utilized in any form or by any means, electronic, or mechanical,
including photocopying, recording or by any information storage and
retrieval system, without written permission from the copyright owner.

Printed in the Netherlands

TABLE OF CONTENTS

FOREWORD

Throughout the Western world, the relationship between gender, science and mathematics has emerged as critical in a variety of contexts. In tertiary institutions, the study of "gender issues", frequently with reference to science and mathematics, is of central significance to many disciplines. Gender studies are being offered as separate courses or parts of existing courses in preservice and postgraduate teacher education, women's studies, technology studies and policy studies. In addition, in the broader context of education at all levels from primary/elementary through to higher, concerned policy-makers and practitioners frequently focus on the interaction of gender, science and mathematics in their attempts to reform and improve education for all students.

In all of these contexts, there is an urgent need for suitable texts, both to provide resources for teachers and students and to inform policy-makers and practitioners. This book has been developed specifically to meet this need. It is designed to be used throughout the world in a variety of tertiary courses and by policy-makers concerned with activities which interface with the gender/science/mathematics relationship. It provides examples which illustrate vividly the rich field from which practitioners and policy-makers in this area now can draw. Its particular appeal will stem from its practical approach and creative future perspective, the international renown of the authors and the generalisability of the recent research and thinking presented in each of the chapters. The comparative analyses of effective programs presented in Section III are the first of their kind in the field and will be of particular interest. I commend this book wholeheartedly to all key audiences.

DOROTHY GABEL
Immediate Past President
National Association for Research in Science Teaching

ACKNOWLEDGEMENTS

The editors wish to acknowledge

Diane Youdell of the Science and Mathematics Education Centre at Curtin University of Technology, for her outstanding support for the production of this book;

The members of the International Gender and Science and Technology (GASAT) community (several of whom are represented as authors or co-authors of chapters) for their support and inspiration over a number of years;

The national Key Centre for Teaching and Research in School Science and Mathematics, for providing the kind of environment which made this work possible.

LESLEY H. PARKER, LÉONIE J. RENNIE AND BARRY J. FRASER

INTRODUCTION AND OVERVIEW

This book builds on two major, world-wide phenomena of recent years. The first
relates to the increasing momentum and variety of studies of the relationship between
gender, science and mathematics. The second phenomenon concerns the growing
tendency for organisations at many different levels (from the national and state levels
through to the small special interest group) to produce policy declarations addressing gender equity, frequently with specific reference to science and mathematics
education.

Two decades ago, there was only a small number of researchers and practitioners
working in the area of gender, science and mathematics. All came from a few Western countries and most were women. Overall, the area struggled to find legitimacy
in the world of international educational research. The majority of studies focussed
on documentation and explanation of girls' apparent avoidance of higher level mathematics and the physical sciences. Generally, explanations were framed from what
Kelly (1986, p. 1) has termed a 'broadly psychologistic angle' implying that girls
needed to change in order to participate fully and achieve well in science and mathematics. Strategies for change generally were associated with the notion of 'non-
sexist' education, which accepted unquestioningly the dominant paradigms of science
and mathematics and, within this context, aimed to provide the same kind of science/
mathematics education for all students.

Today, research in this area is well established. A wide variety of researchers and
practitioners are contributing to the increasing knowledge base in the area. They
come from many different countries and from many different perspectives. Increasingly, these perspectives involve a view that the problem lies not with females themselves, but rather with the nature of science and mathematics and the presentation of
these disciplines in school and society. Increasingly, also, initiatives associated with
these perspectives are known as 'gender-inclusive'. This term appears to have origi-
nated from work in Australia by Jean Blackburn, who argued that educational equality
should be operationalised in ways which 'open up possibilities of a better life for
men and children as well as for women' (1982, p. 16). Within this 'gender-inclusive'
frame of reference, science and mathematics educators now are working to implement an education in which both boys and girls acquire an image and an understanding of science and mathematics which embrace all human beings. In this context,
many of these researchers and practitioners have been associated with the growth of
the second phenomenon which provides a building block for this book, namely, the
proliferation of relevant policies.

Several years ago, certain aspects of this second phenomenon attracted our
attention. To us, it was becoming increasingly evident that a legislative and policy

L.H. Parker et al. (eds.) Gender, Science and Mathematics, xi–xv.
© 1996 *Kluwer Academic Publishers. Printed in the Netherlands.*

mandate for gender equity, although immensely valuable as a legitimising agent, is simply not enough to produce deep-seated changes in educational practice or greater sex equity of outcomes, especially in mathematics and science. The ideas in the policies are fine, but the reality of their implementation is something quite different. In the words of T.S. Eliot (1963, p. 91):

> Between the idea
> And the reality
> . . . Falls the Shadow

In the case of gender and science and mathematics, the 'shadow' appeared to us to be a very long one! Our aim thus became one of shortening the shadow – of bringing the idea and the reality closer together by focussing especially on the various people in education systems who are in positions to effect change.

Our theme in this book is that the key to the translation of gender equity policy into practice lies in working through change agents (for example, teachers, teacher educators and curriculum writers) to ensure that these agents, first, recognise and understand the importance of different world views of science, mathematics, scientists and mathematicians, second, are able to critique current practices in science and mathematics education with respect to gender equity issues and, third, are familiar with the advantages and disadvantages of a range of strategies for addressing gender issues in science and mathematics education. In other words, we focus on perspectives, practices and possibilities. In doing so, we draw together the work of researchers and practitioners based in Australia, Canada, Germany, New Zealand, Norway, the UK and the USA and, in some cases, we highlight the similarities and differences between countries and what can be learned from these. We have given the book a strong future orientation, drawing from recent research and practice examples of strategies and approaches which have been used successfully, and pointing out the ways in which these strategies offer considerable hope for the future.

In many ways, this book represents the first attempt to assemble, in a systematic way, a range of solutions to the problematic relationship between gender, science and mathematics and a theoretical framework which provides a basis for future research and practice in this area, especially in the context of teacher inservice.

Following this introductory section, we have structured the book in three sections focussed, respectively, on perceptions, practices and possibilities.

Section I: Confronting Perceptions and Attitudes contains five chapters focussing on the world views of scientists, pupils, teachers, teacher trainees, curriculum writers and researchers. Each chapter emphasises that effective action to address gender issues in mathematics and science education must be premised on an understanding of these different and often conflicting perspectives. In Chapter 1, *Science in a Masculine Strait-Jacket*, Jan Harding (from the UK) demonstrates that gendered perceptions of persons and of science separate women from science. She argues that the constraining effect of stereotyping must be challenged, so that the world is able to benefit from a simultaneous recognition of the plurality of human beings and the diversity of science.

The other four chapters in this section take up the theme of gendered perceptions focussing, respectively, on preservice teachers (Haggerty), students (Jarvis), teachers and curriculum writers (Willis) and researchers (Johnston and Dunne). Specifically, in Chapter 2, *Towards a Gender-Inclusive Science in Schools: Confronting Student Teachers' Perceptions and Attitudes*, Sharon Haggerty (from Canada) describes action research which aims to encourage the development in science teachers of awareness of their influence on their students' views of science, scientists and science careers. In Chapter 3, *Examining and Extending Young Children's Views of Science and Scientists*, Tina Jarvis (from the UK) discusses the pervasive masculine image of science. She reports her research focussed on young children and is able to trace some of the formative influences on this image. Sue Willis (from Australia), in Chapter 4 (*Gender Justice and the Mathematics Curriculum: Four Perspectives*), discusses four broad views of the relationship between the school mathematics curriculum and gender-based disadvantage and social justice. She describes a way in which a framework which combines these four perspectives can be used with teachers and curriculum writers to provide a starting point for developing a more consistent approach to addressing gender issues in the curriculum.

Finally, Chapter 5, *Revealing Assumptions: Problematising Research on Gender and Mathematics and Science Education*, concludes the section on a reflective note. In this chapter, Jayne Johnston (from Australia) and Mairéad Dunne (from the UK) raise questions about some of the fundamental assumptions of gender research in mathematics/science education. They set a challenging tone for the remainder of the book, reminding us of the necessity of maintaining a critical position, in which we seek constantly to understand how gender is constructed in the research process and in the translation of research findings into practice.

Section II: The Reality of Schools, Classrooms, Curriculum and Assessment comprises six chapters which describe and analyse the current situation in science/mathematics education regarding achievement, assessment, curriculum, pedagogy and classroom interaction. In Chapter 6, *Under Cover of Night: (Re)Gendering Mathematics and Science Education*, Terry Evans (from Australia) provides a link between the perspectives discussed in Section I and the practices which provide the focus for Section II. He emphasises that the public 'daytime' activities, policies and pronouncements of science educators mask a contradictory world of private 'night-time' gendered and gendering processes. He argues that science only can be 'regendered' equitably if scientists and science educators take gender seriously in their public and private lives.

Chapter 7, *Patterns of Science Achievement: International Comparisons* (John Keeves from Australia and Dieter Kotte from Germany), provides us with a powerful reminder that, as recently as 1984, marked patterns of gender differences in science participation and achievement were evident in both developed and developing countries. Keeves and Kotte present evidence of some important differences between countries, and of changes which occurred during the period between the First and Second IEA (International Association for the Evaluation of Educational Achievement)

Science Studies in 1970–1971 and 1983–1984, respectively. They suggest that these changes could well be linked to social and educational programs implemented during this 13-year period to increase the participation and achievement of girls in science.

Chapters 8, 9 and 10 explore specific aspects of the practice of science and mathematics teachers. *Equity in the Mathematics Classroom: Beyond the Rhetoric*, is the focus of the chapter by Gilah Leder (from Australia). On the basis of her research in mathematics classrooms, she suggests that differential classroom treatment and behaviours of females and males are possible sources of gender differences in mathematics learning and, simultaneously, of students' gendered perceptions of themselves as learners of mathematics. Patricia Murphy (from the UK) explores similar issues in relation to assessment. In *Assessment Practices and Gender in Science*, she identifies and discusses sources of gender bias in current science assessment practices. This is followed by a chapter by Kenneth Tobin (from the USA) (*Gender Equity and the Enacted Science Curriculum*) in which he describes sex-differentiated patterns of classroom interaction. He argues that these patterns contribute to females becoming one of the 'involuntary minorities' which view themselves as disenfranchised from the scientific community and do not try to become part of science.

The final chapter in this section is by Jane Butler Kahle (from the USA). In *Equitable Science Education: A Discrepancy Model*, she describes a range of current practices in primary and secondary school science education. She contrasts this actual state with an 'ideal' state, based on her own and others' research of more gender-inclusive schools and classrooms. The chapter thus provides a link to the third section of the book, in which a variety of authors explore possibilities for the future, based on their experiences in implementing gender-inclusive approaches to science and mathematics education.

Section III: From Policy to Practice – Building on Experience contains five chapters describing programs, projects and approaches which have been implemented successfully in several different countries. The writers of the five chapters analyse their experiences. They focus on features which are perceived to be critical to program success, thus providing a foundation for future action. In Chapter 12, *The Role of Persuasive Communicators in Implementing Gender-Equity Initiatives*, Thomas Koballa Jr. (from the USA) addresses the applicability of research on persuasive source characteristics to implementing gender equity initiatives. Central concerns are, first, the legitimacy of persuasion as a form of social influence and, second, the ethical responsibility of the message source.

Chapters 13 and 14 provide practical accounts of two projects – one implemented in Norway and the other in New Zealand. In *Sharing Science: Primary Science for Both Teachers and Pupils*, Doris Jorde and Anne Lea (from Norway) describe the background, methodology and initial outcomes of the *Primary Science Project* developed and implemented under the auspices of the University of Oslo Centre for Science Education. They emphasise that primary science teachers in Norway share many of the problems of their peers in other countries, in that they have little science background and tend to avoid teaching science topics whenever possible. Jorde and

Lea describe how they worked with primary teachers to develop science materials which the teachers felt comfortable teaching and which formed the basis of an enjoyable and effective learning experience for both girls and boys. Bev Farmer (from New Zealand), in *'Do You Know Anyone Who Builds Skyscrapers?' SOS — Skills and Opportunities in Science for Girls*, describes the intervention program, which she developed in collaboration with others to illustrate and promote science-based industrial careers for women. She discusses both successful features and limitations of this program.

The next two three-part chapters describe situations in which closely-related programs have been implemented in both the USA and Australia. Researchers in both countries have collaborated to produce a cross-cultural comparative analysis of the programs, focussing in particular on the manner in which future action and research can build on the findings of the projects described. In *The Politics and Practice of Equity: Experiences from Both Sides of the Pacific*, Nancy Kreinberg (from the USA) and Sue Lewis (from Australia) present a comparative analysis of the strategies used by the EQUALS program in California and the McClintock Collective in Australia. Based on the possibilities and limits revealed by this analysis, they develop a vision for the future. Then, in *Informing Teaching and Research in Science Teaching through Gender Equity Initiatives*, Léonie Rennie and Lesley Parker (from Australia) and Jane Butler Kahle (from the USA) describe and analyse an approach used successfully in both Western Australia and Ohio, USA, in the training of primary school teachers for gender-inclusive teaching of a physical science topic. As part of their discussion, they present a model which provides a framework for future research and action in this area. The chapter and the book conclude with an illustration of the application of this framework, with particular reference to earlier chapters in this volume.

Curtin University of Technology, Perth, Australia

REFERENCES

Blackburn, J. (1982). 'Becoming equally human: Girls and the secondary curriculum', *VISE News* July–August, 16–22.

Eliot, T.S. (1963). *Collected poems of T.S. Eliot (1909–1962)*, London, Faber and Faber.

Kelly, A. (ed.) (1986). *Science for girls?*, Milton Keynes, Open University Press.

SECTION I

CONFRONTING PERCEPTIONS AND ATTITUDES

JAN HARDING

1. SCIENCE IN A MASCULINE STRAIT-JACKET

During the second half of the twentieth century, there has developed a greater willingness to admit that neither gender nor science is absolute or given. Increasingly, both are acknowledged to be social constructs, heavily dependent on cultural contexts, power relationships, value systems, ideological dogma and human emotional needs. A number of writers (e.g., Easlea, 1981; Keller, 1985) have argued that gender is bound up inextricably in the development and practice of science. What is clear to educationists is that interactions between gender and science result in the alienation of women and girls from science, especially from the physical sciences and, in particular, from physics.

In this context, the purpose of this chapter is to present examples of the accumulating evidence about the strength of current gender constructs, and the manner in which these constructs can operate to create and sustain sex differentiation of involvement in science.

GENDER AT WORK

Getting Them Young

A lecturer at Waikato University in New Zealand introduced her Women's Studies course in 1986 with a visit to the local toy shop (Sue Middleton, personal communication). The students were required to place themselves in the position of a visitor from another planet; their brief was to generate a description of the culture of our planet, using only the toys and their packaging as evidence. Of course, the displays in most toy shops are very sexist, separating toys for girls from toys for boys. It takes courage to give a little girl a train set and foolhardiness to give a little boy a dolls' pram.

The way in which day-to-day interactions between a child and its carer can shape its behaviour is shown by an investigation carried out at Sussex University in the UK (Smith & Lloyd, 1978). A number of six-month-old babies each was dressed sometimes as a boy and sometimes as a girl. Mothers who themselves had a six-month-old infant were invited to play with the children in a room containing a variety of toys. It was found that the sex of the baby influenced the mothers' responses, so that the kind of language which they used differed and the same physical action of the baby evoked different responses. If the baby, presented as a boy, was restless, the mother would further activate 'him' with rough and tumble play. On the other hand, if the same child was dressed as a girl, the mother would comfort 'her', sing or read to 'her'.

3

L.H. Parker et al. (eds.) Gender, Science and Mathematics, 3–15.
© 1996 *Kluwer Academic Publishers. Printed in the Netherlands.*

Thus, by the time children are expected to participate in formal school science, they are considerably experienced in gender-appropriate activities. It is of interest, therefore, to examine studies which suggest that certain early activities correlate with later achievement and participation in science.

Early Experiences and School Science: UK Survey Evidence

Between 1980 and 1984, the Assessment of Performance Unit (APU) carried out five annual national surveys in science for ages 11, 13 and 15 in three parts of the UK (England, Wales and Northern Ireland). The surveys were concerned both to assess children's performance in science and to discover some of the reasons for variation in performance. In a report on performance differences between boys and girls, Johnson and Murphy (1986) point out that there was a complex pattern of sex differences in achievement. Girls as a group did better than boys in the sections entitled *Planning Investigations* and *Making and Interpreting Observations*. For *Using Measuring Instruments*, girls' achievement was equal to that of boys, except for specific instruments (microscopes, force meters, ammeters and voltmeters), when boys achieved higher scores overall. At ages 11 and 13 years, boys and girls also performed equally in *Using Graphs, Tables and Charts* but, at age 16, boys' scores on these tasks were higher than those of girls.

It is in the context of physics that girls are reported as showing an overall weakness. This emerged for the APU study for *Interpreting Presented Information* and, very markedly, for *Applying Physics Concepts*. To explore possible causes for the

TABLE I

Differences in the percentages of 11-year-old boys and girls reporting to have 'quite often' engaged in particular activities at home or otherwise out of school (APU survey data 1984)

Activity	% of pupils reporting engagement		
	Boys	Girls	Difference
Make models from a kit	42	6	+36
Play pool, snooker, billiards	59	30	+29
Play with electric toy sets	45	16	+29
Build models using Lego, etc.	50	23	+27
Take things apart to see inside	38	18	+20
Go fishing or pond dipping	30	13	+17
Watch birds	30	27	+3
Sow seeds or grow plants	30	34	−4
Look after small animals/pets	52	57	−5
Collect/look at wild flowers	8	27	−19
Weigh ingredients for cooking	29	60	−31
Sew or knit	5	46	−41

+ Difference in favour of boys.
Based on Johnson and Murphy (1986).

girls' poorer performance, a sub-sample of 11-year-olds (n = 5,000, with equal numbers of boys and girls) was surveyed with a short questionnaire inquiring into engagement in particular activities (Johnson & Murphy, 1986). Clear gender differences emerged as shown in Table I.

It can be seen from Table I that differences in favour of boys occurred in physics-related activities. Moreover, the investigators reported from other studies that: 'Activities involving electricity consistently emerge among those showing the largest differences for boys and girls' (Johnson & Murphy, 1986, p. 20). They explain this observation in the following way:

The extreme *discrepancy* in performance of girls and boys in questions featuring *electricity* . . . is evident across the framework, and is not confined to questions demanding conceptual understanding. Whenever a circuit diagram or an actual circuit is featured in a question girls perform at significantly lower levels than boys; this applies whether they are asked merely to take instrument readings or whether they have to name specific components in a circuit diagram or to follow such a diagram to actually wire up a circuit. (p. 11)

Early Experiences and School Science: Research in Norway

At the University of Oslo in the early 1980s, the Girls and Physics project involved a national survey of 1,200 11-year-olds from 60 schools. The pupils were tested on ability to understand simple concepts and phenomena from everyday experience. Items were grouped into the following categories: technical tasks; comprehension of physics quantities; spatial and geometrical tasks; comprehension of physical phenomena; and reading scales. The results shown in Table II indicate again that girls did not show uniform underachievement across tasks (Lie & Bryhni, 1983, p. 207).

The young people in the study also were offered a list of 27 activities of relevance to science and asked to indicate whether they had experienced these 'rather often', 'once or twice' or 'never'. A number of the activities were similar to those in the APU study and the results, in terms of 'activity rates', were similar. When differences in activity rates were examined in conjunction with differences in test scores,

TABLE II
Mean scores of boys and girls in different categories of tasks as percentage of highest obtainable score

Sex	Mean score					
	Technical tasks	Quantities	Spatial tasks	Phenomena	Reading scales	Total
Boys	89	58	57	54	63	61
Girls	75	56	47	44	60	54
Difference	14*	2	10*	10*	3	7*

*Differences in favour of boys (p < 0.05).
Based on Lie and Bryhni (1983).

the researchers reported an 'amazing correspondence . . . The girls underachieve just in the fields where they have less experience' (Lie & Bryhni, 1983, p. 212).

Again, gender differences in activity and test scores in Norway were greatest in 'electricity', which makes even more significant a project in Western Australia which enabled upper primary girls to achieve parity with boys in assessment of work in electricity carried out with teachers who had undergone a special inservice program (Parker & Rennie, 1985). This program addressed three issues: the development of skills and attitudes related to the teaching of the topic Electricity to grade 5 students; the development of positive teacher attitudes towards the participation of girls in the physical sciences; and the development of skills in creating and maintaining a non-sexist learning environment.

Early Experiences and School Science: The GIST Project

As a result of work carried out in the Girls Into Science and Technology (GIST) project in the UK, Kelly (1987) questions the existence of a strong association between re-ported involvement in science activities and choice of, and success in, science. Part of GIST involved administration of a Scientific Activity questionnaire to pupils in their first term at secondary school. Factor analysis of responses identified three scales: one relating to 'tinkering' items, one to biological activities and one to more theo-retical or academic activities.

At the end of their third year in secondary school the subject examination scores of these pupils and their choice of subjects for future study were recorded. Strong correlations were found between the theoretical scale and subject choice for both boys and girls and between the theoretical scale and achievement for boys, but none of the scales correlated strongly with girls' achievement. However, for girls, the choice of technical craft and achievement in physics correlated highly with 'tinker-ing', indicating some agreement with the other studies reported here.

What's Normal for an Adult?

If we move from young children to adults, we find continued pressure to conform to gender stereotypes. A study carried out in the USA some 20 years ago by Broverman and colleagues is reported by Walum (1977, p. 8). The study involved mental health clinicians – significant professionals in a culture in which consulting an analyst is commonplace. A checklist of 38 items of human personality and behavioural traits was used. Each item was expressed as a dichotomy with a feminine pole and a mas-culine pole, identified through earlier research. Examples of items included 'not at all ambitious' versus 'very ambitious' and 'dislikes mathematics and science very much' versus 'likes mathematics and science'.

The clinicians were divided into three groups. The first group was asked to indi-cate for each item the pole towards which a normal, healthy, mature and competent man would tend; the second group was asked to do the same for a mature, healthy, normal, competent woman; and the third group was asked to evaluate a mature, healthy and competent adult (sex unspecified).

The clinicians were in substantial agreement as to what constitutes a healthy male, female and adult and no significant difference was found between the judgements of male and female clinicians. A normal healthy competent male tended to be seen as independent, objective, unemotional, dominant, competitive, active, skilled in business, self-confident, ambitious, frequently taking the lead and having a liking for mathematics and science. On the other hand, a normal female was considered to be submissive, easily influenced, not adventurous, dependent, subjective, excitable in a crisis, conceited and to have a dislike for mathematics and science. The most disturbing finding was that the 'normal healthy adult' matched exactly the normal healthy male but maturity for women was different; femininity was not defined as adult behaviour.

Perhaps gender perceptions have changed in the USA in the 20 years that have elapsed since this study was reported, but similar stereotypes still can be detected in the UK, where employment shows marked stratification between males and females: few women are in technical jobs and few are in positions of responsibility and decision making. Thus, current gender stereotyping not only ascribes separate activities to girls and boys, but it devalues women and their place in economic and public affairs.

THE MASCULINISATION OF SCIENCE

Kelly (1985) has identified four distinct senses in which science is masculine. One relates to the preponderance of males who choose to study and work within it, creating a visible association of males, rather than females, with science. Another is that science is packaged for learning to suit the ways in which boys are connected to the world, relating to their interests and motivations. The third identifies behaviours generated in science classes, with boys and girls acting out appropriate gender roles, which facilitate learning for boys but restrict it for girls. These three interactions each could make science appear to be masculine, even if it were not. They are also within the power of schools to modify, given the will. The fourth sense is that science is inherently masculine. Its social construction in a patriarchal, or male-dominated, society results in built-in features which actively discourage girls and women from studying it (Manthorpe, 1982).

Keller, who has written extensively in this field, agrees that science is inherently masculine but presents a rather more complex picture. She shows how gender ideology changed in Europe over the centuries and how this led to modern science and the development of an industrialised society:

The scientific revolution . . . did both respond to and provide crucial support for the polarisation of gender required by industrial capitalism. In sympathy with, and even in response to, the growing division between male and female, public and private, work and home, modern science opted for an ever greater polarisation of mind and nature, reason and feeling, objective and subjective, it offered a de-animated, de-sanctified and increasingly mechanised conception of nature. In so doing, modern science gave (at least some) men a new basis for masculine self-esteem and male prowess. (Keller, 1985, pp. 65–66)

However, although modern science led to 'the mastery, control and domination of nature', the values embodied in the underlying ideology never have been accepted universally by the community of working scientists. Keller uses the life and work of Barbara McClintock as a case study to argue that McClintock could work in a different way, with 'a feeling for the organism', not because she was a feminist, but because she was a committed scientist and, being not a man, more readily could take advantage of a pluralistic tradition, avoiding the 'monolithic rhetoric' of masculine science.

Keller maintains that a first step towards a transformation of science would be 'the undermining of the commitment of scientists to the masculinity of their profession that would be the inevitable concomitant of the participation of large numbers of women' (Keller, 1985, p. 175). But perhaps the consequence would be 'inevitable' only if those women are self-reflective and the questioning of the nature of science assumes a greater part than at present of foundation science courses at all levels.

GENDERED PERCEPTIONS WITHIN EDUCATION

Primary Images

Work reported elsewhere in this volume (Jarvis, Chapter 3) has demonstrated the strongly masculine image of science. The GIST project addressed this image. It invited women scientists and technologists into classrooms to teach units of science or technology relating to their work. However, Whyte (1986) reports that, unless the fact that they were female was specifically noted and discussed, many pupils, particularly boys, did not remember that the visitor was a woman. It seems that stereotyped expectations, unless strongly challenged, distort reality.

In Britain, at least one science text is attempting to modify the male imagery of a scientist. Longmans' *Nuffield Science, 11–13*, depicts bespectacled, elderly, white-coated males at a laboratory bench and asks 'Are scientists really like this?' A subsequent illustration show a young woman manipulating technical instrumentation and the text discusses her work as a forensic scientist. Perhaps more stories of women scientists should be available for the primary school.

In the USA, the Career Orientated Modules to Explore Topics in Science (COMETS) project, in action since 1979, has set out to encourage the use of women as role models. COMETS is intended to teach science while, at the same time, allowing students to be exposed to women science career models and to see that science is used in all kinds of careers in their community (Smith, 1987, p. 336). Three kinds of role model are used: the local person (a woman if possible) who shows how she uses in her work the particular piece of science which the class is studying; the historical role model – an account of a woman who worked in the field in the past; and an account of a contemporary young woman scientist who currently works in a science-related field. An evaluation study has demonstrated the effectiveness of COMETS in developing positive attitudes in both boys and girls towards scientists and towards women in science.

Teachers' Perceptions of Pupils as Scientists

Goddard Spear (1987) involved science teachers in England in what was ostensibly an assessment exercise. Teachers were sent samples of work originating from 12-year-old pupils, with a request to assess these samples using a number of criteria and also to predict the potential science future of the child writing them. For some teachers, the work carried boys' names and for others the same pieces of work were headed by girls' names.

The shocking outcome, which science teachers find hard to accept, was that, when the work was labelled as that of a boy, it was graded significantly higher, and the future in science was predicted as a more successful one, in terms of qualifications and employment, than when it carried a girl's name. The same result was found for men and women teachers.

Implications of this study are not that teachers deliberately set out to downgrade work by girls but that, as a group, these teachers did not expect good science work from girls, nor anticipate that they would do significant work as scientists. In an impersonal marking exercise, their judgement was clouded by gender-distorted expectations. One only can speculate about what biases might be displayed in the teachers' day-to-day interactions with young people.

Choice and Perception of Self: Counselling by Computer

The use of computer programs to assist students to make subject choices in schools in the UK could be reinforcing gender stereotyping. This is ironic, because one reason for computer use is to avoid the influence of possible biased advice from a personal counsellor. One such program is JiiGCAL (Job ideas and interests Generator, Computer Assisted Learning). The part for use in subject choice at age 13 years required pupils to indicate which of paired activities they would rather do and how much they like each activity. Computer analysis of responses produced five factors, related to broad 'interest' areas, which were ranked for each individual. Rank 1 indicates the most preferred activity area, and rank 5 the least preferred. Pupils were expected to take note of these rankings when making their subject choices. When it was used for a whole year group in a large comprehensive school, the pattern reported in Table III was recorded.

It appears that the design of the instrument used caused the young people to focus on past experiences and present perceptions of self, which had been generated by sex-stereotyped expectations and did nothing to tap their potential enjoyment of activities not earlier experienced. In this context, it is of interest that a clear factor

TABLE III

Computer feedback to girls and boys of rank of 'interest' area, as percentage of own sex in year group

Interest area	Ranked 1 and 2		Ranked 4 and 5	
	% Boys	% Girls	% Boys	% Girls
Scientific	57	13	27	69
Social	27	82	52	7
Artistic	48	72	26	7
Clerical	27	30	56	29
Practical	42	1	38	85

Based on personal communication from the school's Head of Year 9.

emerging from 'second-chance' courses for women in technology is women's own surprise at their success in, and their enjoyment of, the work which they were doing (Fawcett Society, 1987).

Choice and Job Characteristics

Studies have shown that girls and boys display some differences in their value systems when contemplating the kind of job that they would like to do. The National Child Development Study in Britain has followed the fortunes of all the children born in one week in March, 1956. At age 16 years, they were asked to identify which of 12 characteristics of jobs would be most important for them (Fogelman, 1979). Table IV shows the first five characteristics which emerged for each sex.

Variety, good pay and promotion prospects were important for girls and boys. Boys' preference for thinking or for doing appeared next, while 'helping others' was the most important consideration for one-fifth of girls (only 4% of boys chose this). The convenience of hours of work for girls is likely to assume importance through their concern for home and family commitments.

TABLE IV

Percentage of girls and boys ranking particular job characteristics in first position of importance

Girls (n = 5,818)		Boys (n = 6,042)	
Job characteristic	% Ranking first	Job characteristic	% Ranking first
1. Variety	25.4	1. Well paid	24.0
2. Opportunity to help others	20.3	2. Variety	18.6
3. Well paid	16.9	3. Good promotion prospects	11.9
4. Good promotion prospects	9.4	4. Use of head, thought, etc.	11.8
5. Convenient hours and conditions	7.5	5. Work with hands	11.6
Total	79.5		77.9

An underlying concern for others is evident on the part of the girls. This could be an important motivating factor also in girls' readiness to participate in technology activities. As part of the GATE project (Girls and Technology Education) at Chelsea College, London, the entries of young people in a National Design Prize Competition were analysed. Boys and girls worked on similar devices at age 14 years (the youngest age group and the only one to contain a significant number of girls) but their entries differed in their definition of the problem on which they had worked. The boys generally were attempting to improve a device, but most of the girls located 'the problem' in the social context. They were concerned to help a young child to learn or an elderly or disabled person to be more independent (Grant & Harding, 1987).

These findings demonstrate that girls do not have to become more like boys to be more involved in technology. If curriculum materials in technology are presented to show the relevance of values which girls hold, girls are likely to participate as readily as boys do.

WHO CHOOSES SCIENCE?

Personality, Person Orientation and Choice of Science

Attention has turned in the past decade to questioning the personality of those who choose science. Head's early paper (1979) gives some support for the stereotype of the scientist that children portray. His review of studies carried out in the 1960s generates a profile of a male scientist who is more conservative, authoritarian, hard working and emotionally reticent, and who has a lower person orientation, than his peers.

Smithers and Collings (1981) have investigated the subject choices of able young people in English sixth grades (16- to 18-year-olds). From their initial report that girls choosing physics saw themselves as being less feminine, attractive, sociable and popular than other girls, they went on to use a battery of personality tests with a larger sample, focussing on subject choice groups, but treating boys and girls as separate groups within these. One of these tests was a person orientation scale of 20 Likert-type items, carrying a score varying from 20 to 100; the higher the score, the greater the person-orientation.

Person orientation was defined as 'a preference for dealing with, or involving oneself in, emotional, social or interpersonal issues, as distinct from impersonal ones' (Collings & Smithers 1984, p. 57). The researchers report that: 'Person orientation emerged as an important discriminator in every analysis undertaken and outweighed all other personality constructs in the various multivariate statistical analyses' (Collings & Smithers 1984, p. 62). The person orientation scores for students in different subject groups are shown in Table V. However, although person orientation discriminated the scientist from the non-scientist, girl scientists scored higher on person orientation than boy non-scientists; only girl physical scientists scored lower

TABLE V
Person orientation scores for students by subject group and sex

Sex	Person orientation score				
	Science	Non-science	Mixed subjects	Physical science	Biological science
Boys (n = 787)	57.86	66.49	61.74	55.85	60.64
Girls (n = 1,110)	68.23	76.66	72.44	64.81	69.79

Based on Collings & Smithers (1984).

than the latter. Thus, if science is presented as essentially impersonal, it is likely to be more attractive to the less person orientated among girls as well as boys.

Maturation and Science Choice

Head (1980) investigated maturation and science choice at age 14 years (the age at which choice was offered until recently in English schools). This study showed that a high proportion of boys who had chosen science were among the least mature in the sample, whereas the converse was true for girls. Head comments that girls would need to have a certain maturity to make the unconventional choice of science.

The immature boy scientists displayed emotional reticence combined with a tough rather unfeeling attitude to others. Head suggests that the adolescent boy who has not come to terms with his developing sexuality could find refuge in a science presented as unambiguous, law-bound and 'objective', and remarks that much secondary school science is presented in this way, especially in physics classes.

Object-Relations and Science Choice

Head (1985) and Harding and Sutoris (1987) turned to 'object relations' theory of personal development in an attempt to integrate the various findings of empirical research. They suggested that the way in which we nurture boys and girls generates certain emotional, psychological and cognitive strengths and weaknesses which are relevant to science choice. A child, deprived too early of emotional support, could develop a need to be in control, to suppress ambiguity and to go for certainty, again matching the way in which science, particularly physics, often is presented at the school level. A child without these control-orientated needs could process personal knowledge in a different way, which allows for ambiguity, accepts complexity and does not require abstract generalisations to confer certainty. This child is unlikely to find science, as typically presented in school, an attractive field of study.

Psychological Types and Science Choice

Head and Ramsden (1990) took up the challenge of cognitive style and used the Myers-Briggs Type Inventory (MBTI) to investigate context of science choice. When used with grade 12 students, it was found to discriminate between boy and girl scientists and between girl scientists and girls who had chosen arts subjects. A high proportion of girl scientists fell into a category (described as the SJ quartile) which is characterised by realistic decision making, focussing on immediate experience, seeking an ordered environment and being organised and dependable. When used with 14- and 15-year-old girls, who expressed a commitment to continue with science studies, more than 50% also were allocated to this category.

The test then was used with pupils who had participated in an intervention program at the lower secondary level in which physics was presented as arising out of everyday life and the needs of people. Analysis showed that more of the girls than usual were planning a career in science and these no longer clustered in the 'SJ quartile'. Head and Ramsden (1990, p. 120) commented that 'the experience of 'girl-friendly' physics lessons widens the appeal of science to girls of a greater spectrum of psychological types'.

Much of the physical science curriculum in schools is presented in a depersonalised, abstracted form which attracts to it a certain type of emotionally reticent person, usually male, who has developed needs to control, to abstract and to suppress ambiguity. Not only does this exclude many girls and women, but it constrains the development of science, permitting only certain ways of knowing – the ones in which the chief practitioners are comfortable. There are other dangers, too. Because nurturance, relational responsibility and person orientation are poorly represented in the chief practitioners, these values will not be influential in the development of new science and technologies. Thus, there are created for the planet and its peoples hazards which could be avoided if school science were modified, and the strait-jacket around science itself loosened.

CONCLUSION

In the research cited earlier in this chapter, the clinicians perceived that to be 'female' was to be less than adult. This chapter, however, suggests that to be 'masculine' is conceived to be less than human.

It is difficult to escape from gender. The power of sexual attraction and of procreation renders gender an all-pervading construct in most societies. Unhappily, this all too often is followed by dichotomising other areas of human experience and linking the poles to those of gender. This has happened within the power structures of most societies with the result that activities associated with females possess a lower status and 'interpretations' present the female as deficient. However, because women are involved most frequently with nurturing and caring, the process of dichotomising denies these qualities to men, who then become deficient themselves.

Science also is affected. At a crucial period in the development of modern science, gender constructs interacted to establish its dominant values and ways of working, placing science in a masculine strait-jacket. Keller (1986) appeals for a less gendered science, not just an alternative science which is more 'feminine', but one in which the plurality of science is recognised and permitted. But she argues that it is very difficult to 'count past two' in order to achieve this.

Perhaps the time is ripe for further interactions. Environmental issues are assuming increasing importance and, if the planet is to survive, require inclusion of the nurturant, caring values ascribed to females in the practice of technology. Will this also infiltrate science? Will increasing the number of women in science make a difference? Perhaps it will, if we widen the attraction of science through its presentation, thus facilitating the entry of females and of males with different value systems from the majority who currently choose it. Then perhaps a science which has been released from its masculine strait-jacket will enable humans 'to count past two'.

Gender constructs must be modified. As we set out to facilitate the entry of girls and women into science, mathematics and technology, we must recognise that, unless parallel changes are effected within society, gains could be more apparent than real.

International Gender and Science and Technology (GASAT) Association, UK

REFERENCES

Collings, J. & Smithers, A. (1984). 'Person orientation and science choice', *European Journal of Science Education* (6), 55–65.
Easlea, B. (1981). *Science and sexual oppression*, London, Weidenfeld and Nicholson.
Fawcett Society (1987). *Getting started*, Report of the 1987 Positive Action Awards, London, The Fawcett Society.
Fogelman, K. (1979). 'Educational and career aspirations of sixteen-year-olds', *British Journal of Guidance and Counselling* (7), 42–56.
Goddard Spear, M. (1987). 'The biasing influence of pupil sex in a science marking exercise', in A. Kelly (ed.), *Science for girls?*, Milton Keynes, Open University Press, 46–51.
Grant, M. & Harding, J. (1987). 'Changing the Polarity', *International Journal of Science Education* (9), 335–342.
Harding, J. & Sutoris, M. (1987). 'An Object-Relations Account of the Differential Involvement of Boys and Girls in Science and Technology', in A. Kelly (ed.), *Science for Girls?*, Milton Keynes, Open University Press, 24–36.
Head, J. (1979). 'Personality and the pursuit of science', *Studies in Science Education* (6), 23–44.
Head, J. (1980). 'A model to link personality characteristics to a preference for science', *European Journal of Science Education* (2), 295–300.
Head, J. (1985). *The personal response to science*, Cambridge, Cambridge University Press.
Head, J. & Ramsden, J. (1990). 'Gender, psychological type and science', *International Journal of Science Education* (12), 115–121.
Johnson, S. & Murphy, P. (1986). 'Girls and physics: Reflections on APU survey findings', *APU Occasional Paper 4*, London, Department of Education and Science.
Keller, E.F. (1985). *Reflections on gender and science*, New Haven, CT, Yale University Press.
Keller, E.F. (1986). 'How gender matters: Or why it's so hard for us to count past two' in J. Harding (ed.), *Perspectives on gender and science*, London, UK, Falmer Press, 168–183.

Kelly, A. (1985). 'The construction of masculine science', *British Journal of Sociology of Education* (6), 133–154.

Kelly, A. (1987). 'Does that train set matter?: Scientific hobbies and science achievement and choice', *Contributions to the Fourth GASAT Conference*, Vol. 1, Ann Arbor, MI, University of Michigan, 35–44.

Lie, S. & Bryhni, E. (1983) 'Girls and physics: Attitudes, experiences and underachievement', *Contributions to the Second GASAT Conference*, Vol. 1, University of Oslo, Norway, 202–215.

Manthorpe, C. (1982). 'Men's science or women's science or science, some issues relating to the study of girls' science education', *Studies in Science Education* (9), 65–80.

Parker, L. & Rennie, L. (1985). 'Teacher in-service as an avenue to equality in science and technology education', *Contributions to the Third GASAT Conference*, Chelsea College, University of London, 226–234.

Smith, C. & Lloyd, B.B. (1978). 'Maternal behaviour and perceived sex of infant', *Child Development* (49), 1263–1265.

Smith, W.S. (1987). 'COMETS: Evaluation of a national Women in Science program', *Contributions to the Fourth GASAT Conference*, Vol. 2, Ann Arbor, MI, University of Michigan, 335–342.

Smithers, A. & Collings, J. (1981). 'Girls studying physics in the sixth form', in A. Kelly (ed.), *The missing half: Girls and science education*, Manchester, Manchester University Press, 164–179.

Whyte, J. (1986). *Girls into science and technology*, London, Routledge and Kegan Paul.

Walum, L.R. (1977). *The dynamics of sex and gender: A sociological perspective*, Chicago, Rand McNally.

SHARON M. HAGGERTY

2. TOWARDS A GENDER-INCLUSIVE SCIENCE IN SCHOOLS: CONFRONTING STUDENT TEACHERS' PERCEPTIONS AND ATTITUDES

Cartoons and other popular illustrations of scientists typically portray a bald or shaggy-haired, bespectacled, middle-aged male wearing a laboratory coat and exhibiting some form of bizarre behaviour. Such stereotyped views of science and of scientists can influence the attitudes of our youth as they make decisions about future studies and careers (Chambers, 1983; Koch, 1989). Young people who have a negative view of science and of scientists could be reluctant to choose to study science or plan a career in a science-related field.

A Canadian study (Bateson *et al.*, 1986) indicated that students' attitudes towards science in school and science in society generally were positive, but it is distressing to note that there was a decline in the number of students indicating positive attitudes to science and to careers in science both over time (compared to previous assessments) and by grade (attitudes declined as age increased). This general trend also is evident in Canadian universities where enrolments in science are declining at a time when enrolments in other programs are either stable or increasing.

This chapter addresses views of science in relation to current views of the philosophy of science and science teaching, feminist views of science and an ongoing Canadian study of student teachers' conceptions of science and science teaching. In particular, consideration is given to how these concerns can be addressed in science teacher education so as to encourage science teachers to present and promote a more authentic and gender-inclusive view of science.

PHILOSOPHY OF SCIENCE AND SCIENCE TEACHING

Contemporary philosophy of science views scientific knowledge as socially constructed. Current scientific theories and 'facts' are viewed as the best possible explanations for particular phenomena, given our current understanding. Scientific theories remain useful as long as they both account for what we observe and predict future observations. When that is no longer the case, they are rejected. Few modern scientists or philosophers would hold to a view that the goal of scientific research is to reveal the truth about the universe, a view which Nadeau and Désautels (1984) call 'naive realism'. However, this is precisely how science is viewed by many people (Brush, 1989; Edmondson, 1989). Moreover, textbooks used in preservice science methods courses typically present an 'incomplete and inadequate picture of science' (Abell, 1989, p. 8).

This realist view is often implicit, if not explicit, in relation to how science is taught. Matthews (1990) and Carter (1990) argue that it is impossible to teach science

17

L.H. Parker et al. (eds.) Gender, Science and Mathematics, 17–27.
© 1996 *Kluwer Academic Publishers. Printed in the Netherlands.*

18   S.M. HAGGERTY

without reflecting a particular value system and a philosophy of science. School children come to believe that science is about reality, lacks ambiguity, and involves unequivocal facts. At the same time, they see around them a world that is ambiguous and contradictory and where competing interests might not agree on the necessity of conserving energy, building dams or reducing air pollution. Young people ask: 'What is really true? Why do our scientists and politicians not tell us what the truth is about these matters? Why is science failing humanity?'

Concern for these matters is not new (e.g., see Pope & Gilbert, 1983), but it now is attracting widespread international attention. The 1989 and 1992 international conferences on the History and Philosophy of Science in Science Teaching demonstrated the extent of the concern for the role of the philosophy of science in science teaching. Most science curricula identify an understanding of the nature of science as a major goal of science education. Unfortunately, the importance of this is not recognised widely by science teachers, as demonstrated in Canada, where science teachers ranked it least important of 14 science objectives (Orpwood & Alam, 1981). As Bentley and Garrison (1991) have emphasised, there is a critical need for science teacher education to address the philosophy of science if future science teachers are to develop an adequate understanding of the nature of science.

GENDER-INCLUSIVE SCIENCE

The under-representation of women and girls in science is well known and well documented (e.g., Haggerty, 1991; Kelly, 1978; Scantlebury, 1987). Some researchers have attempted to address this situation by examining the differences between males and females to determine why females do not participate in science. Such an approach appears to assume that female participation will increase if we can make females more like males – that is, females do not participate in science because they are inadequate in some way and this can be remedied by changing females to make them more suited to science. Others argue that it is science that needs to be changed.

Harding (1989) suggests that women can enrich science, noting that:

(M)oral and political insights of the women's movement have inspired social scientists . . . to raise critical questions about the ways traditional researchers have explained gender, sex, and relations within and between the social and natural worlds. (p. 17)

The social sciences have experienced a dramatic broadening in the kinds of research methodologies that are accepted. No longer do we believe that tightly-controlled, experimental studies are necessarily the best way to conduct research. The value of taking a holistic view, rather than isolating variables, has become both recognised and accepted. Feminist researchers have been among the leaders in this movement (e.g., Peterat, 1989; Rosser, 1986; Scott, 1985).

However, even if fields such as education and the social sciences recognise an approach to research that some call 'feminist research', what of the natural sciences? Can there be a thing called 'feminist science'? If feminist science does exist, what does it look like? How does it differ from traditional science? What can it offer to science education, particularly to the science education of girls and women?

Keller has noted that the modern view of science as characteristically masculine dates back to the seventeenth century. When the British Royal Society was founded, its purpose was to establish a 'Masculine Philosophy . . . whereby the mind of Man [sic] may be ennobled with the knowledge of Solid Truths' (Keller 1985, p. 52). This contrasted with the traditional Greek science, which was considered weak and effeminate. Keller (1983) quotes Francis Bacon as describing the then new science as a 'Chaste and lawful marriage between Mind and Nature, the purpose of which was to lead Nature to you with all her children and bind her to your service and make her your slave' (pp. 24–25). This was a view of science that not only represented power and control, but excluded women. It also represented a change in the way of doing science, a change that has dominated scientific endeavour since then. Keller advocates an alternative view of science, exemplified by the work of Barbara McClintock, in which the scientist seeks to know and understand a phenomenon, rather than to dominate and control it. Keller (1983) concludes that the task of feminist scientists is 'to reject a masculine science so that science can continue in its attempts, in its ambitions, to be as free of gender as it is advertised to be' (p. 29).

Others argue for an explicitly feminist science. Helen Longino (1989) has examined these arguments and concludes that, in principle, feminist science is possible. However, she believes that in practice it will not be possible to do science as feminists until there is a change in the social and political context in which science is conducted and reported. Elizabeth Fee (1983) agrees, saying that it is unrealistic to expect a sexist society to produce anything but a sexist science. She claims: 'At this historical moment, what we are developing is not a feminist science, but a feminist critique of existing science' (p. 22).

SCIENCE TEACHING AND VIEWS OF SCIENCE

If we aspire to a less sexist society and a less sexist science, we must be concerned about the science that is taught to our young people in school. One way of influencing school science is through teacher education.

In a study of grade 10 biology teachers and their pupils, Lederman (1986) concluded that science pupils' conceptions of the nature of science were influenced positively by science teachers who modelled an inquiry and problem-solving approach to teaching science. A later report, however, noted that the teaching approaches of experienced teachers were not influenced necessarily by their own conceptions of the nature of science (Lederman & Zeidler, 1987). They concluded that more specific attention had to be devoted to the history and philosophy of science and to specific targeted teaching behaviours in science teacher education.

Brickhouse's (1990) study of three teachers also concluded that teachers' beliefs influence students. She found that the students of one teacher, who had a technical realist view of science, rarely asked questions other than procedural questions.

Conversely, Brickhouse's second teacher believed that scientific theories serve as a basis for solving problems. Her students viewed science as a problem-solving activity. The third teacher, only in his second year of teaching, was inconsistent in his approach to teaching science. He had not yet reconciled his own conflicting beliefs about science or about the impact of institutional constraints on his teaching. Brickhouse concluded that more research is needed into the development of and interaction between teacher beliefs and practice, and into how those beliefs and practice affect students' scientific understanding and participation in science. Overall, studies in this area reveal some inconsistencies about the extent to which teachers' views influence their representation of science. Benson (1989), for example, found that any one teacher typically drew on a variety of philosophical perspectives, depending on the content and the teaching situation. Most studies indicate, however, a need for increased attention to teachers' views of the nature of science. This is one of the major foci for the ongoing project reported in this chapter.

THE PROJECT

The project described here is an ongoing investigation of the development of student teachers' conceptions of science and science teaching and learning (Aguirre *et al.*, 1990). One of the goals of this project is to encourage the development of science teachers who are more aware of their practice and of the implications of that practice for how their pupils come to view science, scientists and science careers. The remainder of this chapter presents views of science held by student teachers who have participated in the project.

The Participants and the Program

All participants have been in one-year post-baccalaureate programs at two Canadian universities enrolled in science teaching methods courses which would qualify them to teach secondary science. The data collection has involved administration of an initial questionnaire in September and follow-up interviews in January and April each year. Altogether, 47 students (22 female and 25 male) have participated in the project in 1989–1992.

Throughout the science methods courses, a social constructivist view of learning and of science has been presented. Students have been assigned readings espousing this view and a number of activities aimed at eliciting and discussing their views of science have been implemented (e.g., see Cobern, 1991). Throughout the course, views or philosophies of science have been referred to and students have been asked to state and defend their views. During practice teaching, students also received feedback encouraging them to identify and confront inconsistencies between their views and their practice. The students have been provided with strategies for reflecting on their practice, and have been encouraged to make that practice more consistent with their views of good teaching and good science.

At the beginning of the program, most of the students demonstrated a view of science that Nussbaum (1989) classified as empiricist/positivist. In its simplest form, the purpose of doing science was considered to be to observe and to explain various phenomena so that we can understand them. This view, referred to as 'blissful empiricism' by Nadeau and Désautels (1984), was expressed by 12 of the 22 women, but only 4 of the 25 men. The remaining 31 students (10 female and 21 male) viewed science as a body of knowledge which is derived by observation and experimentation by using 'the scientific method'. Nadeau and Désautels referred to the view that experimentation is used to prove hypotheses as 'credulous experimentalism'. The gender bias in these views had not been anticipated prior to the study. Edmondson (1989), however, also has reported gender differences in views of science held by introductory college biology students. In her study, the few who held constructivist views of knowledge tended to be female and to be better able to integrate knowledge from different domains. Those who held logical positivist epistemologies were predominantly male and believed strongly that the essence of science consisted of absolutes waiting to be discovered by objective scientific methods.

THE STUDENT TEACHERS' VIEWS OF SCIENCE: ONGOING FINDINGS AND DISCUSSION

The student teachers' views of the nature of science are discussed below under four headings: the social construction of science; consensus; the scientific method; and the role of theory. Each of these is considered briefly to compare the views of the student teachers with current epistemological views. Excerpts from interviews with the students are provided to illustrate their views.

The Social Construction of Science

Wheatley (1991, p. 10) examined constructivism as an epistemological basis for school science and proposed two principles of constructivist theory. First, knowledge is not received passively, but is built up actively by the cognising subject (i.e. communication does not convey meaning, but evokes meaning). Second, the function of cognition is adaptive and serves the organisation of the experiential world, not the discovery of ontological reality. These principles reflect the epistemological view presented in the general science methods courses, a view which is distinctly different from the empiricist/positivist view of science initially expressed by the students participating in the project reported here. In the first year of the project, although several of the students did mention the tentativeness of scientific knowledge, only one (Fred, who held a PhD in science) offered a hint of a constructivist view when asked how scientists go about doing science. Fred's view was that 'science is ever changing . . . science is a way of looking for information . . . science is also invention'.

In subsequent years, the nature of scientific knowledge was addressed more specifically in class discussions and some students began to grapple with their views.

For example, John began the year with a fairly sophisticated view, that science was 'a way of looking at physical reality, a system for explaining what happens around us'. Four months later, he replied, 'I still think it's a process or way of looking at the world. Seeing, trying to explain what goes on, and one of the reasons you want to be able to explain what goes on is so you can predict things'. When asked if his view had changed since the beginning of the program, he replied, 'I think the discussions in class might have broadened my ideas. . . . I tended to think of it in terms like, sort of more or less cut and dried, but it's not quite that cut and dried. I didn't think there were feminist issues in science, but [now] I'm quite aware that there are biases and people interpret things in their own way'.

Ian was still struggling with these ideas near the end of the year, as he talked about how his views of science had changed: 'Well, for instance, like that science is, as opposed to realism, as opposed to being really what it is, what exists, that it's more of a, like man-centred. That the laws and things are generated [i.e., due to generative learning] by us in order to try and explain why things occur, or how things occur. . . . It's tough to think about'.

Most students did not provide a constructivist view of science, even by the end of the year. This seems noteworthy, because almost all students did provide a constructivist view of learning when asked to explain how children learn science. Apparently, they accepted a constructivist view of learning, but not a constructivist view of science and scientific knowledge.

Consensus

Scientists use the criterion of consensus to determine currently-accepted scientific knowledge. Thomas Kuhn's (1970) well-known explication of normal science and revolutionary science is based on this criterion. Again, this view is incompatible with the empiricist/positivist view. Consensus also involves recognition of the tentativeness of knowledge, and we already have noted that several students did mention this characteristic of science. However, they considered that scientific knowledge is tentative until it can be proven, rather than until a better explanation or theory becomes accepted. In Fred's words, 'theory has to be tested and *proven and found right or wrong*. If it's proven it can be accepted by others because it will make sense and it will be logical and intuitive' [italics added]. Carter's (1989) study of preservice science teachers' beliefs about science also found that, generally, students express a realist view of science which precludes the need for consensus. In Carter's study, the one exception to this generalisation, like Fred, had worked as a research scientist.

The Scientific Method

The popular view is that the scientific method guides the work of scientists. This view undoubtedly is reinforced in secondary schools and introductory university science courses, which often require students to use a prescribed format (usually including a hypothesis or problem statement, a description of the methods utilised, observations and a conclusion) when recording practical laboratory activities.

Ryan and Aikenhead (1992, p. 572) examined the views of Canadians who were in their final two years of secondary school. When asked to choose from 10 alternative responses the statement that was closest to their view of the scientific method, 40% of students chose 'questioning, hypothesising, collecting data and concluding'. The second most popular response (13%) was 'testing and retesting – proving something true or false in a valid way'. Only 2% chose the response of 'considering what scientists actually do, there really is no such thing as the scientific method'. Although many students rejected the traditional textbook view of scientific method, 64% of them did feel that there was a definite pattern involved in doing science.

The student teachers in this project tended to concur with this view. When asked how scientists go about doing science, most student teachers identified the classical steps of the scientific method. They did not distinguish between how science actually is done and how it is reported.

Theory in Science

Another often misunderstood aspect of science is the role of theory. Edmondson (1989) found that few introductory biology students understood the role of theories, including 'what theories are, how they change, or how they relate to scientific knowledge as a whole. In many cases, students used the terms 'theory', 'model' and 'hypothesis' interchangeably' (p. 135). Kass and Fensham (1990) interviewed university chemists and secondary school chemistry teachers and identified 15 qualitatively different uses of the term 'theory'. Kass and Fensham (1990) reported a category, explanation/prediction, in which theory has both an explanatory role and a predictive role.

Ryan and Aikenhead (1992) define theory as *'explanations* (often mechanistic and associated with visual representations called models) in which scientists place a high degree of confidence'. In contrast, laws are 'general *descriptions* that enjoy a high degree of scientific confidence' and hypotheses are 'very tentative explanations or descriptions that guide investigations' (p. 571). Sixty-four percent of their subjects believed that hypotheses become theories, which in turn become laws as their proofs become more certain. Laws were seen as having been proven to be true.

The views of the student teachers in this project were very similar to those of students in the above studies. Most regarded theories as what might be termed tentative facts. Blair noted: 'Much of science is theory, which gets tested to make it fact' and Betsy explained: 'Scientists have theories to explain things you're not sure of. Theories are not necessarily facts; they are guidelines; they can change sometimes'. As well as seeing theories as a source of questions or hypotheses, some students saw them as a result of scientific endeavours. Blair noted that, 'most questions begin or end with theory' and 'it [theory] is the end product of observations and assessments regarding the physical world'.

The term 'theory' also was used to represent the formal science view, which was contrasted with how things really are. For example, Ellen noted in her practice teaching that, for many teachers, 'a lot of time [was] spent on just getting the theory out there and into the notebook. [There was] a lot of stress on putting notes on the board'. Carl's initial understanding also seemed to be that theory is a generalisation which explains how something occurs in a theoretical or artificial situation, not in reality. In the course of the year, Carl's view of theory did change. Later he said: 'We propose theories as unverified explanations around which experimentation and testing may be designed. Best possible theories are taught as conceptual knowledge upon which additional learning builds. Theories present a challenge to all those engaging in science to disprove or improve existing understanding'. His revised understanding was consistent with the views of most of the other student teachers. Overall, the student teachers had very naive views of the role of theory in science. This is an issue which should receive more attention in the preparation of science teachers.

Conclusion

This chapter addresses a concern for society's views of science and scientists. Since the seventeenth century, science has been characterised as objective and masculine. It has tended to reflect a realist philosophical view. As Carter (1990) concluded in her discussion of gender and equity issues in science classrooms: 'Rarely do our students develop from science courses a knowledge which they can actually use as a conceptual lens for viewing the world. Instead they learn that science is something not of their world. Perhaps by examining the 'woman problem' as an issue of philosophy we may develop a more powerful framework for thinking about science instruction for everyone' (p. 131). Stereotyped perceptions of science limit the appeal of science for many, especially for girls and women.

At the Fourth International Gender and Science and Technology (GASAT) conference, Mona Dahms (1987) prefaced her remarks by saying:

I am a feminist and feminism is the lens I use when analysing the questions of women in science and technology . . . (W)hen I am using the words female/male or feminine/masculine I am thinking of gender as a *social construction, not* as biological sex. (p. 61)

Modern sociology and philosophy of science contend that both science and gender are socially constructed, and are influenced deeply by their social, cultural and political environments (Harding, 1989). Science has been constructed primarily by men and embodies an intrinsically masculine world view (Kelly, 1985). It has been dominated by themes of power, by claims of being objective, and by a technical and positivist view. Feminist scientists argue that bringing a more holistic and personal approach to science will enrich science and make it both more accessible and more valid.

Our work in science teacher education is aimed at modifying how science is taught in schools. As Lesley Parker (1987) has noted:

What is needed is a school science which reflects the diversity of scientific practice not the maleness of much scientific rhetoric. . . . Science needs to be portrayed as the totally human activity that it is in reality. (p. 91)

Similarly, Bentley and Watts (1986) argue the need for some fundamental changes in the way in which science is presented in schools. The changes involve a recasting of the mould of science as objective and dispassionate, a reassessment of the nature of evidence, explanation and their relationship to one another, and a review of the status of scientific knowledge, especially in terms of philosophical and psychosocial aspects of learning environments.

Concerns such as these about how science is represented in schools have been ignored largely in most science teacher education programs. These are issues which must be addressed if we truly are concerned about increasing the participation of women in science.

ACKNOWLEDGEMENTS

The study discussed in this chapter was funded by the Social Sciences and Humanities Research Council of Canada.

University of Western Ontario, London, Canada

REFERENCES

Abell, S.K. (1989). 'The nature of science as portrayed to preservice elementary teachers via methods textbooks', in D.E. Herget (ed.), *The history and philosophy of science in science teaching*, Tallahassee, Florida State University, 1–14.

Aguirre, J., Haggerty, S. & Linder, C. (1990). 'Student teachers' conceptions of science, teaching and learning: A case study in preservice science education', *International Journal of Science Education* (11), 381–390.

Bateson, D., Anderson, J., Dale, T., McConnell, V. & Rutherford, C. (1986). *British Columbia science assessment, 1986: General report*, Victoria, BC., Queen's Printer.

Benson, G.D. (1989). 'The misrepresentation of science by philosophers and teachers of science', *Synthese* (80), 107–119.

Bentley, M.L. & Garrison, J.W. (1991). 'The role of philosophy of science in science teacher education', *Journal of Science Teacher Education* (2), 67–71.

Bentley, D. & Watts, M. (1986). 'Courting the positive virtues: A case for feminist science', *European Journal of Science Education* (8), 121–134.

Brickhouse, N.W. (1990). 'Teachers' beliefs about the nature of science and their relationship to classroom practice', *Journal of Teacher Education* (41), 53–62.

Brush, S.G. (1989). 'History of science and science education', *Interchange* (20), 60–70.

Carter, C. (1989). 'Scientific knowledge, school science, and socialization into science: Issues in teacher education', in D.E. Herget (ed.), *The history and philosophy of science in science teaching*, Tallahassee, Florida State University, 41–51.

Carter, C. (1990). 'Gender and equity issues in science classrooms: Values and curricular discourse', in D.E. Herget (ed.), *More history and philosophy of science in science teaching*, Tallahassee, Florida State University, 122–132.

Chambers, D.W. (1983). 'Stereotypic images of the scientist: The draw-a-scientist test', *Science Education* (67), 255–265.

Cobern, W.W. (1991). 'Introducing teachers to the philosophy of science: The card exchange', *Journal of Science Teacher Education* (2), 45–47.

Dahms, M. (1987). 'Theme presentation: Evaluation', in J.B. Kahle, J.Z. Daniels & J. Harding (eds.), *Proceedings, Fourth (GASAT) Conference*, Ann Arbor, MI, University of Michigan, 48–61.

Edmondson, K.M. (1989). 'College students conceptions of the nature of scientific knowledge', in D.E. Herget (ed.), *The history and philosophy of science in science teaching*, Tallahassee, Florida State University, 132–142.

Fee, E. (1983). 'Women's nature and scientific objectivity', in M. Lowe & R. Hubbard (eds.), *Woman's nature*, New York, Pergamon Press, 9–27.

Haggerty, S.M. (1991). 'Gender and school science: Achievement and participation in Canada', *Alberta Journal of Educational Research* (37), 193–206.

Harding, S. (1989). 'Is there a feminist method?' in N. Tuana (ed.), *Feminism and science*, Bloomington, IN, Indiana University Press, 17–32.

Kass, H. & Fensham, P.J. (1990, June). *Conceptions of theory in chemistry among university and secondary school teachers*, paper presented to the annual meeting of the Canadian Society for the Study of Education, Victoria, BC.

Keller, E.F. (1983). 'Is science male?' in H.L. Ching (ed.), *Proceedings of the First National Conference for Women in Science and Technology*, Vancouver, Society for Canadian Women in Science and Technology, 21–29.

Keller, E.F. (1985). *Reflections on gender and science*, New Haven, Yale University Press.

Kelly, A. (1978). *Girls and science: An international study of sex differences in school science achievement*, International Association for the Evaluation of Educational Achievement, IEA Monograph Studies No. 9, Stockholm, Almqvist & Wiksell.

Kelly, A. (1985). 'The construction of masculine science', *British Journal of Sociology of Education* (6), 133–154.

Koch, J. (1989). 'Educating the educators', in I. Ravina & Y. Rom (eds.), *Contributions to the Fifth GASAT Conference*, Vol. I, Haifa, Technion-Israel Institute of Technology, 70–75.

Kuhn, R. (1970). *The structure of scientific revolutions*, 2nd ed., Chicago, University of Chicago Press.

Lederman, N.G. (1986). 'Relating teaching behavior and classroom climate to changes in students' conceptions of the nature of science', *Science Education* (70), 3–19.

Lederman, N.G. & Zeidler, D.L. (1987). 'Science teachers' conceptions of the nature of science: Do they really influence teaching behavior?', *Science Education* (71), 721–734.

Longino, H.E. (1989). 'Can there be a feminist science?' in N. Tuana (ed.), *Feminism and science*, Bloomington, IN, Indiana University Press, 45–57.

Matthews, M.R. (1990). 'History, philosophy and science teaching – what can be done in an undergraduate course?', *Studies in Philosophy and Education* (10), 93–97.

Nadeau, R. & Désautels, J. (1984). *Epistemology and the teaching of science*, Ottawa, Science Council of Canada.

Nussbaum, J. (1989). 'Classroom conceptual change: Philosophical perspectives', *International Journal of Science Education* (11), 530–540.

Orpwood, G.W.F. & Alam, I. (1984). *Science education in Canadian schools. Vol. II, Statistical database for Canadian science education*, Ottawa, Science Council of Canada.

Parker, L. (1987). 'The reclamation of science: Towards diversity', in J.B. Kahle, J.Z., Daniels & J. Harding (eds.), *Proceedings, Fourth GASAT Conference*, Ann Arbor, MI, University of Michigan, 85–92.

Peterat, L. (1989). 'Re-search and re-form: A feminist perspective in home economics research', in F. Hultgren & D. Coomer (eds.), *Home economics teacher education: Alternative modes of inquiry in home economics research*, Peoria, IL, Glencoe Publishing, 211–219.

Pope, M. & Gilbert, J. (1983). 'Personal experience and the construction of knowledge in science', *Science Education* (67), 193–203.

Rosser, S. (1986). *Teaching science from a feminist perspective*, New York, Pergamon Press.

Ryan, A.G. & Aikenhead, G.S. (1992). 'Students' preconceptions about the epistemology of science', *Science Education* (76), 559–580.

Scantlebury, K. (1987). 'Female enrolment trends in science courses at tertiary institutions, 1975–1985, in B.J. Fraser & G.J. Giddings (eds.), *Gender issues in science education*, Perth, Australia, Curtin University of Technology, 19–29.

Scott, S. (1985). 'Feminist research and qualitative methods: A discussion of some of the issues', in R.G. Burgess (ed.), *Issues in educational research: Qualitative methods*, London, Falmer Press, 67–85.

Wheatley, G.H. (1991). 'Constructivist perspectives on science and mathematics learning', *Science Education* (75), 9–21.

TINA JARVIS

3. EXAMINING AND EXTENDING YOUNG CHILDREN'S VIEWS OF SCIENCE AND SCIENTISTS

In order to take full responsibility in this increasingly technological and scientific world, citizens need to understand the nature of science so that their decisions are informed appropriately. In addition to young children understanding the nature of science, it is also important that every child appreciates that he or she could aspire to a scientific career.

There is much documented evidence that fewer girls than boys choose to study the physical sciences in secondary schools in Western countries (ASE, 1990; Kahle, 1985; Kelly, 1987) and that differences in attitudes and aptitudes that can influence secondary school subject choice are already significant in primary schools (APU, 1988; Ormerod & Wood, 1983; Taber, 1991). Research also indicates that the majority of young children hold narrow stereotypical views of scientists, as white, middle-aged males who work in laboratories (Chambers, 1983; Maoldomhnaigh & Hunt, 1988). Consequently, many girls and non-white children might not even consider that science could be an appropriate career for them. Although children's pre-conceptions, once established, are known to be very tenacious and difficult to change (Ausubel, 1968), it is of interest nevertheless to establish the extent to which focussed activities, for very young children, particularly girls, can have a broadening effect.

This chapter reports a two-part study which, first, explored how young children's views of science and scientists vary with respect to age and gender and, second, investigated the effectiveness of three different intervention strategies aimed at broadening children's views of science and scientists.

METHOD

Six schools in Leicester, UK, were selected to represent the different cultural and social groups represented in this community. One hundred and thirty-four children, aged 5 to 11 years, were involved in the study, including 64 Caucasians, 61 Asians from Indian, Pakistani and African backgrounds, five from Afro-Caribbean cultures and four children of mixed race. There were equal numbers of boys and girls. The collection of data and delivery of teaching activities were carried out by 34 postgraduate teacher trainees. Every trainee worked with about four children representing the ability range in each class, for at least one hour a week over seven weeks.

29

L.H. Parker et al. (eds.) Gender, Science and Mathematics, 29–40.
© 1996 *Kluwer Academic Publishers. Printed in the Netherlands.*

Part One: Assessing the Children's Views

In the first part of the study, three activities were used to assess the children's views of science and scientists. In the first activity, the children were asked to draw pictures of a scientist at work and to describe what they thought science was and what a scientist looked like. The drawings were analysed using the categories developed by Chambers (1983). This 'Draw a Scientist Test' (DAST) already has been used widely (Mason *et al.*, 1991) and enables this study to be placed into the context of other comparable research. Most children were asked to produce two pictures, because some children appear to record what they feel to be the socially acceptable answer first and then record a wider view in their second illustration (Maoldomhnaigh & Hunt, 1988). Although Schibeci and Sorensen (1983) suggest that the DAST gives an accurate representation of the images of scientists held by children, it was felt that the second and third activities needed to be used to probe the children's views further.

In the second activity, the children in the sample were interviewed using photographs showing men, women, people with disabilities and those from different cultures at work. The children were required to select the scientists and give their reasons both for selection and rejection of the pictures. Each set of photographs included the categories described in Table I.

The third activity involved interviews with the children, using pictures of teenage boys and girls, including those with disabilities and from different cultures. There were five pictures of teenagers on one page: a Caucasian boy painting, a Caucasian girl looking at a magazine, a West Indian boy with no activity, a Caucasian boy in wheel chair playing pool, and an Asian girl playing an electronic keyboard. After having been told that all the teenagers shown wanted to be scientists when they grew up, the children were asked to comment on whether these individuals would be

TABLE I

Categories of scientists and non-scientists represented in photographs used in the second activity

Scientists	Non scientists
1. Typical chemical and physics laboratories with males present	1. Males and females wearing uniforms that distinguish the occupation (e.g., nurses and fire fighters)
2. Typical chemical and physics laboratories with females present (e.g., a pregnant scientist lining up a laser)	2. Males and females where an obvious activity is shown (e.g., fishing and picking tea)
3. 'Non-stereotypical' situations/locations with males only present (e.g., agricultural chemists collecting data in a field)	3. Males and females without uniforms and an activity not particularly clear or familiar to young children (e.g., a business meeting and recording a radio program)
4. Non-stereotypical situations with females, disabled scientists or those from other cultures present (e.g., two Asian botanists examining plant growth in an outside location and a woman in a wheelchair at a laboratory bench)	

likely to achieve their ambitions, what they might need to do at school, and whether this was an appropriate ambition at all.

Part Two: The Intervention

There are many suggestions for intervention aimed at increasing the interest and involvement of girls in science. These include organising single-sex groupings (Morgan, 1989), increasing teachers' skills in teaching science and awareness of gender issues (Parker & Rennie, 1986), linking science and technology to provide a social context (Smail, 1984), and monitoring the availability of materials and access to equipment to ensure equality of experience. The few studies of the science-related disadvantage of non-Caucasian children in England, and of ways to optimise these children's access to science, indicate the importance of taking account of the children's language abilities and putting science activities in a culturally-relevant context (Jarvis, 1991; Peacock, 1991).

The study reported in this chapter examined the relative effectiveness of three strategies: use of non-fictional accounts of female and non-Western scientists; creative oral or written response based on science experiments; and cooperative group work. Children throughout the age range had the opportunity of being involved in each strategy. The pupils worked in small groups for least five sessions of one hour's duration. After the final session, the children were asked to draw two more pictures of scientists, and to sort a completely new set of photographs, to ascertain whether intervention had had any effect.

RESULTS AND DISCUSSION OF ASSESSMENT OF CHILDREN'S VIEWS

Tables II and III summarise children's responses to the DAST. The younger children (5- and 6-year-olds) either drew a generalised person without any distinguishing features or a picture that appealed to them, such as a house or a playground (defined as self interest in Table II). The exception in this age group was a group of children who had undertaken activities in a designated 'Science' room. They claimed that they knew what science was and many drew themselves carrying out activities with sand and water. These children appeared to have been able to link their investigative activities with the term science.

The middle years, between ages 6 and 8, were characterised by a mixture of responses which included a very generalised person without any gender or activity indicated, a teacher, an artist and a white male who was very similar to the stereotypical view outlined by Chambers (1983). The drawings by the children closely reflected their comments made during the interviews.

Two interesting alternative models of the 'scientist' (i.e., the teacher and the artist) occurred fairly frequently throughout these middle age ranges. The confusion between the artist and scientist appears to be widespread. The author is aware of this image being produced in several other Leicestershire schools and in some schools in

TABLE II
Types of responses to the request to 'Draw a Scientist'

Type of response	Percentage of children responding at age						
	5	6	7	8	9	10	11
Self interest	64	33	9	9			
Person	12	24	35	12			
Self portrait	12	9					
Teacher	6	19	9				
Artist	0	5	4	12			
Scientist	0	5	39	58	93	100	100
Other	6	5	4	9	7		
Number of children	17	21	23	33	30	8	2

Perth, Western Australia. It appears to relate to the fact that scientific skills of observation are developed through accurate drawings and because young children communicate many of their scientific findings in the form of pictures, paintings and models.

From about the age of 8 years, the 'stereotypical' view of the scientist increasingly was established. Initially the children concentrated on the appearance and clothes which closely match the image identified by Chambers (1983), with the exception that very few children drew facial hair, although the hair itself often was depicted as eccentric in some way. The most common piece of equipment, according to the children, were eyeglasses, with a laboratory coat being almost as important. The

TABLE III
Number of categories from DAST recorded by children in each drawing of recognisable scientists

Number of categories depicted in each drawing[a]	Percentages of children responding at age					
	6	7	8	9	10	11
One	100	33	5			
Two		23	37	7	12	
Three		33	21	46	38	
Four		11	32	18	50	100
Five			5	7		
Six				14		
Seven				4		
Eight				4		
Number of children who drew scientists	1	9	19	28	8	2

[a]DAST Categories: male, white, laboratory coat, eyeglasses, facial hair, symbols of research, symbols of knowledge, products of technology and captions.

older children included more symbols of research, such as bottles, test tubes and microscopes, with the 10- and 11-year-olds occasionally including symbols of knowledge and relevant captions (see Table III).

These findings are similar to those identified by Newton and Newton (1992) who administered the DAST to children aged from 4+ to 11+ years in Great Britain, where they found that the stereotypical view tended to form as early as 6 years of age and was complete by the time children reached 11 years. Solomon (1993) identified at least four distinct images that were common with children aged 9 to 14 years: the weird, male individuals engaged in dangerous chemical activities; the authoritative and helpful doctor and teacher who tests and explains things for us; the technologist who makes artefacts, tests to see if they work and improves them; and the scientist who mostly is interested in ideas and designing experiments to see if predictions work. Solomon found that only the helpful teacher/doctor model was age-related, and common with the younger children in this group.

All these models were identified in the 7- to 11-year-olds in this sample but with a different distribution. Although the children aged from 8 to 11 years depicted a white male chemist involved in science experiments, the 'weird' element was missing in most pictures, except for the oldest primary children. In addition in this study, there was a tendency for the doctor, teacher and technologist to occur more in the 7 and 8 years age group rather than throughout the age range, with the addition of the 'artist-scientist', not identified by Solomon. These differences could be an indication that more children in this study had the experience of being taught science as part of the implementation of a National Curriculum in their early years.

It is of interest that, despite the fact that half of the children were not Caucasians, there were only seven coloured scientists drawn, of whom six were drawn by children from Asian cultures and one by a Caucasian in her second drawing. Similarly there were few female scientists depicted. Only five female scientists were drawn in the first pictures and an additional four children drew women in their second drawings, with all but one of these being drawn by girls. The finding that few women scientists are drawn by children, and that the women scientists usually are drawn by girls, also conforms with other research (Chambers, 1983; Kahle, 1989; Maoldomhnaigh & Hunt, 1988).

Selection of Photographs of Scientists

In the second activity, the greatest value of the photographs proved not to be the numerical results relating to the children's choice, but the reasons for selection of scientists. Children's reactions to the photographs indicated that some very young children had some concept of a scientist before it showed in their drawings and, despite the fact that the children's drawing showed an overwhelming number of white male scientists, the children did not use gender at all in choosing the photographs of scientists. On only one occasion was a picture rejected on the grounds of race. This indicates that the narrow stereotypes depicted in the children's drawings are not entirely representative of their understanding of scientists, but it seems probable that these limited images, nevertheless, are likely to have an effect on the children's attitudes.

Several children had incorrect models which they applied consistently, as in the case of a 5-year-old who consistently picked out every picture for which she felt that the adult was helping, such as a police officer assisting children to cross the road, and lifeboat personnel helping drowning people. Many of the youngest children chose scientists on the grounds that they were wearing a white coat, glasses and/or gloves, even when the person shown was not involved in a science activity, such as a woman working in a shoe factory. Scientists were often rejected on the grounds that this apparel was absent, or for example, 'because scientists don't wear shorts'.

Most groups of children aged 8 years and over appeared to be able to consider more than one factor and weigh up their decisions, whereas the younger children tended to use one factor per picture. The older children not only used dress, but also used the presence or absence of typical laboratory equipment, computers and electrical equipment as significant. Several groups rejected photographs where the adult was shown to be working outside. All but one of the 7 to 11 years age group used investigating as another deciding factor. This clearly is an important idea to build on, to give a realistic, wide and non-stereotypical view of science.

Comment on Teenage Aspirations to Become Scientists

Table IV provides a summary of the children's responses to photographs of teenagers whom they were told aspired to become scientists. Again, despite the narrow gender and race orientation of the children's own drawings, when they came to comment on whether teenagers could become scientists, these factors were mentioned rarely. Instead, children of all ages overwhelmingly concentrated on what the teenagers were doing. Disability was seen as a significant factor by some children. The picture of the boy in a wheelchair provoked much discussion. Many of those who gave reasons for rejecting him as a potential scientist felt that he would be unable to escape from explosions or go out to collect data.

Overall, many children felt that potentially anyone could be a scientist, but that it required hard work and ability. Science as an occupation was perceived as very serious and single-minded and not associated with enjoyment at all. A number of teenagers were rejected as potential scientists because they were not perceived to be working hard or were playing. The Afro-Caribbean often was rejected because he was looking happy or not doing anything, although a few children chose him as they thought he was smiling because he had completed all his work. Kahle and Lakes (1983) reported that, when responding to questions concerning science as a career choice, 13- and 17-year-old girls felt that working in science would not be fun or would be too much work. This view appears to be prevalent even with very young children, possibly inhibiting them from choosing this career later.

TABLE IV
Children's opinions about which teenagers are likely to achieve their ambitions to be a scientist

Description of photograph	Number of children		
	Accepted	Rejected	No opinion
Caucasian boy painting	78	5	28
Caucasian girl reading a comic	75	15	21
Afro-Caribbean boy	49	40	22
Caucasian snooker player in a wheelchair	47	41	23
Asian girl playing an electronic keyboard	59	27	25

In summary, the results from the three instruments indicate that the youngest children have little or no idea of what science is or have an incorrect concept. A rather narrow view of the scientist as a serious, isolated white male working in a laboratory develops as early as six years in age. However, it should be pointed out that children's concepts, although limited, are usually wider than indicated by their drawings alone.

RESULTS AND DISCUSSION OF THE INTERVENTION

Non-Fictional Accounts of Female and Non-Western Scientists

The importance of a historical element has been recognised in secondary science as promoting better learning of the concepts of science, increased interest and motivation, a better attitude of the public towards science and an understanding of its social relevance. For example, Solomon et al. (1992) reported that middle-school pupils (aged 11–14 years) showed some areas of substantial progress in their understanding of the nature of science when this element was included. Additionally, Smail (1984) suggests that the inclusion of biographical material about scientists can make science more interesting to girls, because it builds on their interest in people, shows science as a social activity and scientists with relationships with others.

In this study, it was decided to use stories covering the lives and work of past women and non-Caucasian scientists. The teacher trainees presented the stories orally, adjusting the content to take into account the age and experience of the listeners by using pictures of the scientists, puppets and simple drama. In addition to hearing the story, children over 8 years old were prompted to discuss general issues. For example, they were encouraged to realise that there were not many black and women scientists in the past because they did not have the educational, financial and social opportunities that white males had at that time. The majority of the children responded well to these discussions. There was general agreement among the children that discrimination on the grounds of race was unjust, whereas there was persistently more disagreement between the boys and girls about the importance of a woman's occupation.

TABLE V

Percentage of children in each age group drawing scientists before and after the non-fiction intervention

Occasion of drawing	Percentage of students at age					
	5	6	7	8	9	10
Before intervention	0	0	50	30	78	100
After intervention	86	17	67	60	100	100

Many boys considered that mothers should be at home with their children rather than following careers as scientists.

As shown in Table V, after intervention, all age groups recorded more recognisable scientists, with many pictures showing a slight increase in detail of equipment, indicating the children's increasing knowledge of materials used by scientists. It was obvious that the children had been influenced strongly by the accounts, because several pictures showed scenes from the stories. In the 7–10 years age range, there also was a noticeable reduction in males drawn as scientists, with a total of 13 women scientists depicted, several recognisable from the stories. All these pictures were drawn by girls. Two coloured scientists were drawn, one by a 5-year-old Asian girl and the other by a Caucasian girl. As Maoldomhnaigh and Hunt (1988) found, girls appear to be more prepared to extend their view of scientists.

Although the approach of using non-fictional stories of scientists had some effect with all the children, it was more successful in terms of science understanding with children over 7 years of age. As much science work in the early years is presented through general topics that cover many curricular areas without the actual science being specifically identified, many children struggled to link the stories to scientific activity. Therefore, a prerequisite for this type of approach is for teachers to tell the children which activities are science in their day-to-day activities so that the children can build up a concept of science and its processes.

Creative Oral or Written Response Based on Science Experiments

With the intention of finding ways of increasing the numbers of females aspiring to and entering scientific careers, Smail (1984) advocated the use of imaginative writing as a 'girl-friendly' aid to assimilating scientific principles and ideas. Martinez (1992) also investigated modifications of science experiments including introducing a fantasy scenario and placing activities into a social context. He found that in general such interest enhancements were more effective for girls than boys.

In the study reported here, children were asked to create oral or written fictional stories, with themselves in the role of the hero/scientist, based on technological and scientific problems, which had been investigated practically in the classroom. One group of 5-year-olds used the properties of mirrors to save a 'friend' trapped in a castle by a monster and a group of 7-year-olds tested materials to help wrap up a lost kitten which had been left on a cold mountain-side.

This approach was time consuming, as more than an hour per session was required so that the science investigation could be developed fully. In addition, the method appeared to have little effect on the children's views of scientists as depicted in their drawings. Even after eight weeks, 10 children virtually redrew their original drawings, replicating even fairly minor details. This did not occur with the other two strategies. There were some changes in some of the older children's responses. A few commented that they could have drawn anyone as they had been doing science themselves. Although this approach seems to have minimal value in widening children's view of science, there were indications that it had other value, especially for the younger children, because it provided a clear context and helped to remind them of the science concepts.

Cooperative Group Work

Research in Great Britain and USA indicates that cooperative group work in general promotes social skills, self-esteem, discussion, testing, inferring and giving informed conclusions, thus providing a classroom atmosphere which is preferred by most girls and many boys (Aronson et al., 1978; Cowie & Rudduck, 1988; Dunne & Bennett, 1990; Kahle 1989). There is also research that indicates that, when children who hold different scientific concepts are grouped together to carry out a collaborative task, the resultant discussion and negotiation encourages the children to confront their limited perceptions and make adjustments (Howe, 1990). It appears that cooperative grouping can both enable children to articulate their views of science and widen them in discussion with their peers, as well as being a strategy that would appeal to a wide range of children.

The final intervention strategy, therefore, was to give groups an activity which required one agreed outcome. The groups were asked first to evaluate a non-fiction science book, taking into account whether boys, girls, non-white and disabled children were likely to identify with the illustrations and activities. Using these discussions, the group then was required to produce its own illustrated booklet on scientific activities that would appeal to all children.

Although the books produced by the groups were of a high standard and showed a wide range of science activities depicting men and women working both indoors and outside, this extended view was not reflected in many of the final pictures of scientists. This could indicate that children need time and further activities to establish this wider view. There were some changes in that the model of the teacher as scientist was lost and there was a small increase in the number of women depicted (four as opposed to two before intervention). Additionally, the pictures of scientists showed fewer of the stereotypical features identified by Chambers despite the fact that pictures were very detailed and showed considerable care and commitment, thus indicating that the older age range had developed a less stereotypical view. One of these instances was the second boy who recorded a female scientist during the study, the first boy having copied a girl's picture. This 9-year-old boy drew a very detailed picture and said that he deliberately was drawing a woman without a laboratory coat.

This chapter has presented evidence of children's images of science and scientists and evaluated three strategies aimed at modifying these images. With respect to children's images of scientists, it was found that the only 5-year-olds who felt able to draw a scientist before intervention were those who had been told when they were doing science, albeit in a narrow context. By the end of the study, far more of the children were able to provide an appropriate drawing of a scientist and were clearer about what science was. Consequently, it is recommended that teachers make it clear when children are doing science activities, even when they are incorporated within cross-curricular activities.

With respect to the intervention strategies, it was found that the stories of past scientists were enjoyed by children of all ages and appeared most effective in extending the children's view of science and scientists. However, the very young children were most interested in the personal details of the people, rather than the actual scientific issues or historical perspective. One might argue that as 5- and 6-year-olds have not developed any view of science, let alone a stereotypical view, it is unnecessary to introduce the idea of women and non-Caucasian scientists at all. However, both this study and others indicate that the stereotypical view is beginning to develop from about this age. In addition, a number of misconceptions, such as the 'artist-scientist', develop at the same time. Therefore, it could be more appropriate for younger children to be presented with stories of modern-day scientists involved in familiar business or industrial enterprises with an emphasis on their day-to-day lives. Historical accounts and the opportunity to participate in discussions of social and racial issues, linked to both ongoing science and history activities, then could be most appropriately introduced to children aged over 7 or 8 years.

Although there are indications that a strategy which encourages creative responses from children has value for developing literacy skills and increasing motivation and recall of science concepts, the evidence from this study suggests that it had limited effect in widening the children's view of scientists in the short term. The group work strategy had some effect in enabling the children to develop a wider view of a scientist, as shown by an increased number of scientists or science activities drawn, and the reduced number of stereotypical descriptions of scientists used by older children. Although all the children participated willingly in the activities, cooperation was less effective among the younger children. Cooperative group work skills require considerable time to develop and the intervention appeared to be more effective in those classes where the children already were used to some cooperative work.

An appreciation of science as an investigatory process is a prerequisite for a wide, non-stereotypical understanding of scientists. Science for the primary school as advocated by the National Curriculum in both England and Australia, with emphasis on developing the skills of observation, testing, predicting and hypothesising in the context of science concepts, has the potential for achieving this among young children. However, unless educators have a clear understanding of the ways in which children's images of science and scientists evolve, the new curricula will have limited effects in increasing the children's appreciation of science in general.

Leicester University, UK

REFERENCES

Aronson, E., Blaney, N., Stephan, C., Sikes, J. & Snapp, M. (1978). *The jigsaw classroom*, London, Sage.

Assessment of Performance Unit (APU) (1988). *Science at age 11: A review of APU survey findings 1980–1984*, London, Department of Education and Science/Her Majesty's Stationery Office.

Association for Science Education (ASE) Gender Issues Working Party (1990). *Gender issues in science education: Occasional papers*, Hatfield, ASE.

Ausubel, D. (1968). *Educational psychology: A cognitive view*, New York, Holt, Rinehart and Winston.

Chambers, D. (1983). 'Stereotypical images of the scientist: The draw-a-scientist test', *Science Education* (67), 255–265.

Cowie, H. & Rudduck, J. (1988). *Co-operative group work: An overview*, Sheffield, BP Educational Service.

Dunne, E. & Bennett, N. (1990). *Talking and learning in groups*, Basingstoke and London, Macmillan.

Howe, C. (1990). 'Grouping children for effective learning in science', *Primary Science Review* (13), 26–27.

Jarvis, T. (1991). 'Science and bilingual children: Realising the opportunities', *Education 3–13* 19(1), 41–48.

Kahle, J. (1985). *Women in science: A report from the field*, New York, Falmer Press.

Kahle, J. (1989). *Images of scientists: Gender issues in science classrooms*, (What Research Says to the Science and Mathematics Teacher 4), Perth, Australia, Curtin University.

Kahle, J. & Lakes, M. (1983). 'The myth of equality in science classrooms', *Journal of Research in Science Teaching* (20), 131–140.

Kelly, A. (ed.) (1987). *Science for girls*, Milton Keynes, Open University Press.

Maoldomhnaigh, M. & Hunt, A. (1988). 'Some factors affecting the image of the scientist drawn by older primary school pupils', *Research in Science and Technological Education* (6), 159–166.

Martinez, M. (1992). 'Interest enhancements to science experiments: Interactions with student gender', *Journal of Research in Science Teaching* (29), 169–177.

Mason, C., Kahle, J. & Gardner, A. (1991). 'Draw-a-scientist test: Future implications', *School Science and Mathematics* (91), 193–198.

Morgan, V. (1989). 'Primary science – gender differences in pupil responses', *Education 3–13*, 17(2), 33–37.

Newton, D. & Newton, L. (1992). 'Young children's perceptions of science and the scientist', *International Journal of Science Education* (14), 331–348.

Ormerod, M. & Wood, C. (1983). 'A comparative study of three methods of measuring the attitude to science of 10 and 11 year-old pupils', *European Journal of Science Education* (5), 77–86.

Parker, L. & Rennie, L. (1986). 'Sex-stereotyped attitudes about science: Can they be changed?', *European Journal in Science Education* (8), 173–183.

Peacock, A. (1991). *Science in primary schools: The multicultural dimension*, Basingstoke, Macmillan Educational.

Schibeci, R. & Sorensen, I. (1983). 'Elementary school children's perceptions of scientists', *School Science and Mathematics* (83), 14–20.

Smail, B. (1984). *Girl-friendly science: Avoiding sex bias in the curriculum*, York, Longman.

Solomon, J. (1993). *Teaching science, technology and society*, Buckingham, Open University Press.

Solomon, J., Duveen, J. & Scot, L. (1992). 'Teaching about the nature of science through history: Action research in the classroom', *Journal of Research in Science Teaching* (29), 409–421.

Taber, K. (1991). 'Gender differences in science preferences on starting secondary school', *Research in Science & Technological Education* (9), 245–251.

4. GENDER JUSTICE AND THE MATHEMATICS CURRICULUM:
FOUR PERSPECTIVES

The past two decades have seen a rapid increase in the attention paid to gender issues in school mathematics. This has occurred at the level of policy, research and practice. In this period, we have learned a lot and much has been achieved. Gender differences in achievement and participation in mathematics no longer are regarded as either natural or inevitable. Curriculum materials and assessment tasks are, at least, less overtly sexist and, indeed, they are often more consciously inclusive of what are perceived to be the experiences and concerns of girls. Many mathematics classrooms are considerably more 'friendly' to girls than they once were – both in regard to the general pedagogical approach and in the way in which girls are treated. Many girls and boys are happy to assert that 'everything is equal now'. Often their teachers and members of the general public also believe this, or even that the balance in education has swung in favour of girls and to the detriment of boys.

In a volume such as this, it hardly needs to be said that everything is *not* equal now – no more in school mathematics than in life generally. Notwithstanding the academic successes of many girls and young women in school mathematics, it would be difficult to deny that school mathematics continues to be gender-inflected (see, for example, Dowling, 1991; Fennema & Leder, 1993; Jungwirth, 1991; Walkerdine, 1989) and that many mathematically oriented disciplines and occupations remain predominantly male domains even if they no longer are almost exclusively male domains. We still have a long way to go.

Over the past seven years, I have been involved in research, curriculum development and professional development activities which have informed my understanding of how gender equity is understood by those involved in mathematics curriculum development and delivery. The purpose of this chapter is to outline some of the differences in perspective which I have observed regarding the relationship between the school mathematics curriculum, gender-related educational disadvantage and social justice.

GENDER EQUITY AND THE SCHOOL MATHEMATICS CURRICULUM

As part of a long term research project, colleagues and I undertook case studies of gender equity initiatives of secondary schools across Australia (Kenway & Willis, 1993; Kenway et al., 1991). While this project did not focus specifically on mathematics, in the schools in which we worked the lower retention of girls than boys in mathematics and related fields was known well, commented upon widely and, very often, seen as a major gender issue to be addressed. Nevertheless, curriculum reforms

L.H. Parker et al. (eds.) Gender, Science and Mathematics, 41–51.
© 1996 *Kluwer Academic Publishers. Printed in the Netherlands.*

informed by considerations of gender were unusual and only a minority of teachers of mathematics considered that their subjects might be gender-inflected or that reform was either necessary or possible (Kenway & Willis, 1993; Willis, 1995a). The 'problem of retention' was seen to lie with the poor choices which girls make as a result of social and peer group pressure, home influence or lack of accurate information about their options and futures. Its solution was seen as lying elsewhere than in their classrooms. Thus, while teachers endorsed programs directed at encouraging more girls to do mathematics, most were of the view that the curriculum was appropriate 'for all students' or, if it wasn't, that the curriculum reflected the inherent 'nature of the subject' and that any change would be at the expense of 'standards' or 'rigour'. Most believed also that girls and boys had equal access to mathematics courses and were treated equally within them, and that treating students in the same way is what equity is about. There were, of course, some teachers who did not hold these views and such teachers often faced the rather daunting task of trying to bring about curriculum change in an environment ranging from bored disinterest to downright hostility. It is not surprising, then, that school-based efforts to address issues of gender in school mathematics have been sporadic, superficial and unsystematic.

In addition to these research projects, I have been involved closely in two collaborative projects of the States and Territories of Australia to develop, firstly, a national curriculum framework for school mathematics and, secondly, a national profile for documenting students' achievements in mathematics as they proceed through school. These projects occurred during and after a period in which programs were funded by Australian governments and industry in an attempt to increase the participation of girls and women in mathematics; television advertising campaigns exhorted girls to take more mathematics; many education systems and schools developed policies and practices in support of gender equity; feminist (and non-feminist) researchers documented the many ways in which girls have been disadvantaged in or by school mathematics and suggested some of the ways in which change might be brought about; and mathematics teachers undertook a wide range of projects to address girls' lower participation, compared to that of boys in mathematics. In this context, there was some reason to suppose that issues of gender, and of social justice more generally, would inform the curriculum process in which we were engaged. To a certain extent, this was so.

Most people who were involved in some way in the two national mathematics curriculum projects, whether as participants or commentators, claimed to believe in equity and probably sincerely so. As we know, however, there are varying interpretations of what constitutes equity. Does it mean, for instance, equal opportunity to learn mathematics or equal educational treatment or equal educational outcome? (Fennema, 1993). I indicated earlier that, for many mathematics teachers, it means one of the first two, although there are some for whom it means the third. In the process of developing the mathematics documents, differences in perspectives on what equity is, and of what it implies for the curriculum, became a source of some considerable frustration For example, the following paragraph appeared in various drafts:

Children and adolescents will vary in personal interest in mathematics and the extent to which they value it. *When these differences occur along gender, social-class or ethnic lines, the implication is that the mathematics curriculum is not equitable in meeting the needs of different groupings of the community.* Whether a particular student gains the full benefit from mathematics may be influenced by a range of personal characteristics and circumstances. It will also depend on the qualities of the mathematics offered. [italics added]

Regularly, in consultation feedback, it would be suggested that the sentence in italics be changed to read:

When the mathematics curriculum is not equitable in meeting the needs of different groupings of the community, differences will occur along gender, social-class or ethnic lines.

When asked to elaborate, without exception those who proposed reversing the sentence believed that the revised version was what the writers had been trying to say. While the writers were coming from the position that equity for social groups was about equality of outcomes for social groups, these particular respondents were coming from one of the other two positions on equity listed above. Importantly, they had not recognised these different perspectives and believed the original version to be a more clumsily written version of their own (as one commentator said 'why don't you say what you mean instead of that convoluted educationalise'). Indeed, when the document went through a formal editing process, the sentence was changed in three successive edits before the writers convinced the editor that the meaning had been changed and that the altered meaning was not the intended one.

As I suggested earlier, there is a widespread belief in the need to increase the participation of girls and women in mathematics and related fields, whether for reasons of social justice (that is, for the good of the girls and women), or to better serve the needs of the others (that is, for the good of 'the economy', industry, government, higher education or the broader community) (see Mura, 1995). Furthermore, addressing 'the needs of educationally disadvantaged groups', one of which was defined to be girls, was part of the brief for the curriculum projects in mathematics. Nevertheless, throughout the writing and consultation processes, it became clear that there were widely disparate views about what relationship, if any, the mathematics curriculum has to gender justice and, consequently, about whether and how the mathematics curriculum should change (see Willis, 1995b).

Whether at national consultative meetings or in the school staffroom, dealing with differences in viewpoint was hampered severely because participants did not share a common framework or language and hence had difficulty recognising, let alone understanding, each others' points of view. Often the same words meant quite different things to people, with unfortunate consequences in loss of faith when an agreement of one meeting appeared to be broken by the next. Equally often, choices of words that caused offence masked an underlying commonality of viewpoint which eventually would become apparent, although not without considerable stress.

While feminist researchers and gender workers in education might have a deep understanding of the various theoretical perspectives which seek to explain the relationship of the mathematics curriculum to gender inequality, many of those involved in mathematics curriculum development and implementation – whether at the level of national reference groups or at the level of daily classroom practice – do not.

As a result of these research and curriculum development experiences, each spanning over six years, I tried to develop a framework which would enable us to view, to consider and to critique various strategies for addressing gender differences through the mathematics curriculum. In doing so, I identified four broad perspectives on the relationship between the mathematics curriculum, 'disadvantage' and social justice.

THE MATHEMATICS CURRICULUM AND 'DISADVANTAGE'

Within the first of these perspectives, the school mathematics curriculum is taken more or less as a given. This includes what is to be learnt, how it is taught, and how it is assessed. The curriculum might change somewhat over time in order to reflect changing views about what content is needed, and what pedagogy and assessment are most appropriate, but these changes are based on decisions of what is appropriate for the 'universal' student or for 'typical' children. While students can experience different curricula, perhaps based on 'ability' or 'future needs', this is not overtly based on, for example, students' gender or race. Within this perspective, however, some social groups of students are seen to be disadvantaged because, as a group, they are likely to be less willing, well prepared or able than other social groups to learn mathematics.

Within this first perspective, the curriculum is 'innocent' of any role in producing disadvantage or injustice, the problem (which is not to say the *fault*) lies with the children who, because of their gender, race, ethnicity, social class or disability, lack the knowledge, skills or motivations necessary for access to, and success in, school mathematics. In the case of girls, they are alleged to lack certain spatial skills, or to avoid taking risks or to lack the motivation to succeed in mathematics. The solution to this problem is to help such children become better prepared for school mathematics and thus gain the benefits which accrue to those who learn mathematics (that is, participate) and learn it well (that is, achieve). The task of schools and teachers is to provide the children with what they are missing, to ensure that they are in a better position to gain access to, and succeed in, school mathematics. The focus of attention or intervention, then, is the child who will need assistance to overcome her or his disadvantage or readiness for mathematics. For girls, this might involve special workshops to help them develop certain spatial skills, or motivational talks by visiting role models. As a shorthand, I call this the *remedial* perspective (although some consider it is a *deficit* perspective on the disadvantaged learner).

Within the second perspective, we take what mathematics is to be learnt and the order in which it is learnt as more or less fixed or not particularly related to group disadvantage or difference. However, we accept that how it is taught and how it is assessed could be to the relative advantage or disadvantage of some social group-ings of children. Because of the classroom management strategies and pedagogical practices adopted, and the contexts and experiences drawn upon in teaching and assessing the mathematics, children in some social groupings are in a worse posi-tion, either to learn the required mathematics or to demonstrate that they have learned the required mathematics, than children in other groups.

Within this perspective, while the intended curriculum (in content and sequence) could be innocent, the curriculum actually experienced by children is not. The prob-lem is that groups of children do not have an equal opportunity to achieve in math-ematics. This might be because they are treated differently, for example, girls and boys could once have been guided into different mathematics courses which pro-vided them with different opportunities to learn certain mathematics, or within the same classroom they might receive differing amounts of teacher time or different types of questions. It also could be because learning experiences and assessments do not draw equally upon, or relate to, the experiences or learning styles of children from different social groupings. For example, the contexts in which mathematics is embedded could more often be familiar to boys than girls. The solution is to im-prove our classroom practice, textbooks and assessments in school mathematics so that all social groups of children are treated fairly in ensuring true equity of access both to the mathematics that they need to learn and to the means of demonstrating their learning. The educational task is to improve curriculum implementation so that we draw upon equally and extend children's experiences, provide an equally sup-portive learning environment and more valid and fair assessment opportunities. Again, simply as a shorthand, I call this the *non-discriminatory* perspective (in the case of gender, the *non-sexist* perspective).

Within the third perspective, the school mathematics curriculum is regarded as a selection from all possible curricula. We understand that the choices made in devel-oping school mathematics curricula will reflect the values, priorities and lifestyles of the dominant culture and, indeed, the more powerful members of the dominant culture. The choice of content and the sequence also will match the developmental sequences typically associated with the children in the dominant cultural group. Children in non-dominant social groups are forced to learn mathematics which is less consistent with their experiences, interests, cultural practices and developmen-tal sequences, and this will be so even given the best of the pedagogy and assess-ment practices suggested within the second perspective. In this way, the school mathematics curriculum will work in the interests of the dominant social group and, in all likelihood, produce advantage relative to other social groupings.

Thus, with respect to gender, school mathematics could privilege characteristics of mathematics which are identified more closely with the masculine over character-istics more closely identified with the feminine, such as the logical over the intui-tive, the context-free over the context-bounded, the rational and abstract over the personal and social, the unambiguous over the ambiguous, or the absolute over the relative (e.g., Brown, 1984). At a more obvious level, curriculum for 'more able' students in the senior secondary years might assume implicitly an ideal student who is more oriented to the physical rather than the biological or social sciences. Consequently, such a curriculum might privilege certain topics over others, perhaps calculus over statistics. Thus, for some young women, to choose to do the higher level mathematics courses also could be to choose to learn the mathematics least connected or appropriate to their experiences, interests and future needs. Conversely, to choose to do the mathematics most obviously connected to their future needs, and so on, also is to choose to do the mathematics which challenges and stretches them least and which is least valued publicly.

Within this perspective, the problem lies with curriculum content and sequence. The solution is to rethink who school mathematics is for, what school mathematics is, what should be learned, by whom and when, and hence to improve our curriculum development practices in school mathematics so that mathematics curricula reflect better a broad range of social groupings and, equally, work in the best interests of all social groupings of children. The educational task is to provide children with a curriculum which gives value and validity to their own knowledge and experience and which acknowledges and respects diversity and difference between social groups. I call this the *inclusive* perspective.

Within the fourth perspective, the school mathematics curriculum is considered to be implicated actively in producing and reproducing inequality. School mathematics is regarded as one of the ways in which dominant cultural values and group interests are maintained. School mathematics is seen to be linked inextricably to the maintenance of privilege of some groups over others. Through its content and practices, it constructs the successful mathematics learner as middle class and male. Within this perspective, representations of males and females in mathematics textbooks are considered to be rooted more deeply, and removed less easily, than common sense views of sex stereotyping suggest. Rather, they are seen as constructing the mathematics learner as a gendered subject, where gender is a classification which relates to the domestic division of labour (Dowling, 1991). Similarly, through a range of classroom practices, girls and boys (and their parents and teachers) come to believe that girls' successes in mathematics are of a less worthy kind than boys, being based more on hard work, conscientiousness and rule following than on real understanding or 'brains' (Walkerdine, 1989).

Within this fourth perspective, the problem is considered to lie with the mathematics curriculum as a whole and the way it positions, classifies and selects students both inside and outside schools. The solution is to challenge and hence to modify this hegemony, where challenging hegemony requires that it be recognised by the participants. Only in this way can the mathematics curriculum truly work in the services of improved social justice. The educational task is to reconstruct our own views and our students' views of who does mathematics and what it means to be good at it. Whereas those adopting the third perspective would focus on our getting the curriculum content, sequence and pedagogy 'right', those adopting the fourth perspective would suggest that it is never possible to 'get it right', albeit we can get

TABLE I
Disadvantage and the mathematics curriculum

	Perspective 1 Remedial	Perspective 2 Non-discriminatory	Perspective 3 Inclusive	Perspective 4 Socially critical
The mathematics curriculum is . . .	a given, including what is to be learnt, how it is taught and how it is assessed	a given with respect what is to be learnt, but how it is taught and how it is assessed are not	a selection from all possible curricula and therefore neither given nor unchangeable	actively implicated in producing and reproducing social inequality and in being one of the ways in which dominant cultural values and group interests are maintained
The problem of 'disadvantage' lies with . . .	the children, some of whom by virtue of their race, ethnicity, gender, social class or disability are less well prepared than others to get the full benefits of the curriculum	pedagogy and assessment practices which favour or relate to the experiences, interests and cultural practices of some social groupings of children more than others	curriculum content and sequence which reflect the values, priorities and lifestyles of the dominant culture and match the typical develop-mental sequences associated with their children	the way the mathematics learner is constructed through the curriculum and the way mathematics is used inside and outside schools to support and produce privilege
The solution is to . . .	help such children become better prepared for school mathematics	change pedagogy and assessment practices to ensure children have real equity of access both to the math-ematics and to the means of demon-strating their learning	rethink who 'the typical child' is for whom our curriculum is developed, what school math-ematics is, what should be learned, by whom and when	challenge and modify the hegemony of mathematics and use mathematics explicitly in the services of social justice
The educational task is to . . .	provide children with the missing skills, experiences, knowledge, attitudes or motivations	draw upon and extend children's experiences, provide a supportive learning environ-ment and more valid assessment opportunities	provide children with curricula which better acknowledge, ac-commodate, value and reflect their own and their social groups' experiences, in-terests and needs	help children develop different views of who does mathematics and what it means to be good at it, to understand how to use it in the interests of social justice

it 'more right'. Therefore, they will wish to assist students to understand how they and others are positioned by school mathematics and to decide what they want to do about it, and how to use mathematics in their own interests and in the interests of social justice. I call this the *socially critical* perspective.

These four perspectives are summarised in Table I.

USING THE FOUR PERSPECTIVES

I have found these four broad perspectives useful as a framework for engaging teachers and curriculum writers in discussions about the mathematics curriculum, social justice generally and gender justice specifically. It enables them to understand better the positions from which they are coming and where their own curriculum practices are located, and gives groups a starting point for developing a more consistent approach to issues of gender justice and the curriculum. It provides also a means by which movements in perspective can occur and by which contradictory practices can be understood and addressed. Let me illustrate this with two examples.

For over 20 years, writers, illustrators and publishers have been exhorted to improve textbooks to make them less overtly sexist, by removing gender stereotypes and by drawing equally on experiences and interests associated with the 'masculine' and the 'feminine'. Over the past year or so, in professional development workshops, I have asked teachers to analyse the teaching materials which they commonly use. Most start the exercise with a rather 'ho hum' attitude. After all, we all know about sexism in old textbooks – all of that has been fixed. Very quickly they change their minds. For example, in one Australian textbook series:

almost all women (except teachers) [are] referred to as mum or grandma or the butcher's wife, almost always cooking or shopping, and rarely given a name. Men, on the other hand, are occasionally called dad . . . but are mostly given names and described in terms of their job. Men have and handle more money. . . . While mathematics is now regarded as important for all students, perhaps lower levels of mathematics are sufficient for women's work. (Willis, 1995c, pp. 270–271)

One reaction of teachers to the evidence of the deep gender bias of many textbooks is to blame the 'perpetrators' who will have to 'do better'. Another is to castigate themselves for not having noticed. These reactions are located within the second 'non-sexist' perspective. Implicitly, they hold that 'getting it right' is possible if only each of us was more committed or conscientious. However, that the publishers of many of these books have made an obvious effort to remove sexism and that most teachers report having not noticed any problems in the books until directed to study them specifically, suggests just how deeply embedded are the gendered beliefs which lead to these textbooks.

Teachers involved in professional development activities have found it helpful to think about the implications of the socially critical perspective for their understanding of how textbooks still manage to 'get it wrong' after all the efforts which have been made to improve them. Often they come to see that, while there is still room for improvement in textbooks, the level of vigilance required to 'get them right', even if we all agreed about what that meant, would make it a near impossible task. Just how many 'gender police' can we afford? In any case, we cannot protect our students from the same forces outside our classrooms.

At this stage, it is possible to introduce some of the thinking and the strategies used in English classrooms to assist students to develop alternative reading practices. These strategies are based, in part, on the view that this kind of subtle sexism does the most damage when it goes unnoticed and that, rather than focussing all of our efforts on to trying to remove it, we should help students to recognise, understand and deal with it. In mathematics, students could use their data collection and analysis skills to study how women and men are constructed within their mathematics textbooks and to consider how their own feelings about mathematics could be influenced by these representations. They then might extend their investigation beyond textbooks to consider what happens in their own classroom and in their families that might influence their views about gender and mathematics. The possibilities for investigation are considerable.

To move on to a second example, many feminist teachers feel a certain amount of conflict about trying to persuade their female students into current forms of mathematics. Carolyn, a teacher in a girls-only school commented:

There is a greater problem solving emphasis in senior secondary school mathematics. I have continually found students who I know have the conceptual and algebraic skills to successfully complete a problem not even start . . . They seem not to be able to take the risk to make the first few mistakes . . .

My research is not over and this is where I too have a block, a block of a different sort . . . no matter how hard I try to cut my concerns back to a small manageable size I am constantly aware of the broad issues of contradictions between courses', exam's and society's image of mathematics . . . I feel no surprise at them not wanting to take risks in a subject that seems intent in turning them into failures. (Kenway & Willis, 1993, p. 52)

On the one hand, the strategies that Carolyn adopts suggest a remedial perspective, in which the problem to be dealt with is that her girls lack important risk-taking qualities. On the other hand, her 'remedy' feels dishonest, as though she is cheating or setting her girls up.

Part of the problem for this teacher is that she is trying to grapple with the contradictions which she has identified without the theoretical tools with which to do so. Carolyn might find that Walkerdine's (1989) analysis, of the way in which males and females are positioned differently with respect to school mathematics, provides her with some ways of understanding her own unease, her 'block'. It also could provide Carolyn with some ways forward. Walkerdine's work suggests that Carolyn cannot remove the contradictions for the girls whom she teaches, and that it is not only she who needs access to this powerful knowledge, but also her students need to understand how their attributions and their choices about mathematics are socially constructed and constrained. *They* need the knowledge and the tools to understand the nature of gender and power relationships and how mathematics (with other school subjects) is implicated in perpetuating inequalities in these relationships. But, of course, if they have developed coping strategies which are likely to be unproductive for their future learning of mathematics (such as avoiding risks), then they also could need help to overcome this. However, this help will not be in the spirit of 'getting them right', but rather of enabling them to take charge; it will not be done *to them, but rather with* them.

As with any attempt to categorise complex phenomena, this description of four perspectives oversimplifies and hence, to some extent, distorts practice. Each of these perspectives represents a wide range of specific positions, practices and strategies regarding the mathematics curriculum. Importantly, while they involve contradictory positions on the relationship of the mathematics curriculum to disadvantage and social justice (you cannot, at the one time, hold the view that the curriculum is implicated in producing inequality and, at the same time, that is not!), strategies suggested by one perspective could be consistent with another perspective. In this sense, the four do not represent a set of discrete categories. Thus, you might take a socially critical perspective and yet, at the same time, recognise that, because of the way in which the curriculum currently is constructed, some strategies located within a remedial perspective can be warranted for some children (such as by helping them more readily take risks in mathematics).

While my major use for this framework has been in working with teachers and curriculum developers, I have found it useful also in undertaking a preliminary analysis of the data which my colleagues and I have collected in schools. Locating particular teachers' views of gender and the mathematics curriculum within these broad perspectives has proved helpful in understanding their practices, their concerns and their personal conflicts. It has proved also to be a means for identifying the source of conflicts within schools amongst teachers who purport to share a common goal of gender equity in mathematics education. For research purposes, however, this is a rather blunt instrument, which undoubtedly will need considerable revision and refinement.

CONCLUSION

In this chapter I have described four broad perspectives on the relationship of the school mathematics curriculum to disadvantage and social justice, with particular reference to disadvantage associated with gender and the promotion of gender justice. I have not attempted to provide a critique of these perspectives or to argue the case for one over the other (see, for example, McLeod, 1990; Willis, 1995a; Yates, 1993). My intention is rather to provide an overview of the main emphases, arguments and implications for curriculum of each perspective. As I suggested earlier, feminist researchers and gender workers are likely to have an understanding of the various theoretical positions which inform work in gender and mathematics and of the sometimes subtle, but nevertheless significant, differences between them. This is not likely to be the case for the majority of participants in mathematics curriculum review, development or delivery. Whether their work involves them in serving on curriculum committees, in writing curricula and teaching materials, or in teaching students, a framework such as this can assist practitioners to understand, compare and evaluate various strategies for addressing gender differences in school mathematics. It can provide also a vehicle for the development of better understanding between participants in curriculum development and delivery.

Murdoch University, Perth, Australia

REFERENCES

Brown, S. (1984). 'The logic of problem generation: From morality and solving to de-posing and rebellion', *For the Learning of Mathematics* 4(1), 9–2.

Dowling, P. (1991). 'Gender, class, and subjectivity in mathematics: A critique of humpty dumpty', *For the Learning of Mathematics* 11(1), 2–8.

Fennema, E. (1993). 'Justice, equity and mathematics education', in E. Fennema and G. Leder (eds.) *Mathematics and gender*, Brisbane, University of Queensland Press, 1–9.

Fennema, E. & Leder, G. (eds.) (1993). *Mathematics and gender*, Brisbane, University of Queensland Press.

Jungwirth, H. (1991). 'Interaction and gender – findings of a micro-ethnographical approach to classroom discourse', *Educational Studies in Mathematics* (22), 263–284.

Kenway, J. & Willis, S. (with the Education of Girls Unit of South Australia) (1993). *Telling tales: Girls and schools changing their ways*, Canberra, Department of Employment, Education and Training.

Kenway J., Willis, S., Blackmore J. & Rennie L. (1991). 'Studies of reception of gender reforms in schools', in L. Rennie, L. Parker and G. Hildebrand (eds.), *Action for equity: The second decade*, Proceedings of the Sixth International GASAT Conference, Key Centre for School Science and Mathematics, Perth, Australia, Curtin University of Technology, 111–128.

McLeod, J. (1990, December). 'Gender-inclusive curriculum: Some issues for professional development', Paper presented at the Conference of the Australian Association for Research in Education, Sydney, Australia.

Mura, R. (1995). 'Feminist theories underlying strategies for redressing gender imbalance in mathematics', in P. Rogers and G. Kaiser (eds.), *Equity in mathematics education: Influences of feminism and culture*, London, Falmer Press, 155–162.

Walkerdine, V. (1989). *Counting girls out*, London, Virago Press.

Willis, S. (1995a). 'Gender reform through school mathematics', in P. Rogers and G. Kaiser (eds.), *Equity in mathematics education: Influences of feminism and culture*, London, Falmer Press, 186–200.

Willis, S. (1995b). 'The mathematics profile: Removing the flaws', in C. Collins (ed.), *Curriculum stocktake: Evaluating school curriculum change*, Canberra, Australian College of Education, 172–191.

Willis, S. (1995c). 'Mathematics: From constructing privilege to deconstructing myths', in J. Gaskell and J. Willinsky (eds.), *Gender in/forms curriculum: From enrichment to transformation*, New York, Teachers College Press, 262–284.

Yates, L. (1993). 'The education of girls: Policy research and the question of gender', *Australian Education Review No. 35*, Melbourne, Australian Council of Education Research.

JAYNE JOHNSTON[1] AND MAIRÉAD DUNNE[2]

5. REVEALING ASSUMPTIONS: PROBLEMATISING RESEARCH ON GENDER AND MATHEMATICS AND SCIENCE EDUCATION

Over the past three decades, teachers, researchers and policy makers have become increasingly aware of differences in the participation and achievement of girls and boys in school mathematics and science. Significant amounts of research have focussed on this issue, and associated initiatives have been developed, usually with the explicit aim of increasing the participation of girls in these subjects. This work has contributed to our understanding of the area and has affected the educational and career opportunities of some girls. However, we believe that the scope and extent of these changes have been circumscribed by a limited conceptualisation of gender.

The purpose of this chapter is to raise questions about gender research in mathematics/science education, by questioning explicitly some of the assumptions underpinning its research base. In particular, we are interested in considering how gender is conceptualised within this research. Our approach is based on the theoretical framework developed by Habermas (1972). Habermas describes three 'knowledge-constitutive interests' – technical, practical and emancipatory – which foreground the political basis of epistemological positions. Each interest is seen as shaping what is considered to constitute knowledge, and determining the ways in which knowledge is organised socially.

Although Habermas' framework did not arise directly from educational considerations, it provides an analytic tool for interrogating educational practices and research (Grundy, 1987). We note that one of the limitations of using a categorical system is that the description of categories might render them discrete and apparently mutually exclusive. No field of research fits neatly into such categories. Nevertheless, the explanatory power of the categorical system as a tool for interrogating the gender research base outweighs its limitations. The power of Habermas' framework as an analytic tool stems from the position of critique which explicitly constitutes the third interest, the emancipatory. For our purposes, this interest can be understood as constituting a position from which to critique the absence of an explicit recognition of the politics of knowledge constitution in the technical and practical positions.

THE TECHNICAL AND PRACTICAL INTERESTS

At the heart of Habermas' technical interest is the need for control and prediction of the environment. What counts as knowledge is seen as generated through observation and verified through experimentation. This view conveys an image of an external, objective body of knowledge, where the territory and procedures for investigation

L.H. Parker et al. (eds.) Gender, Science and Mathematics, 53–63.
© 1996 *Kluwer Academic Publishers. Printed in the Netherlands.*

are demarcated strongly. There is a strong connection between the technical interest, the scientific tradition and positivist philosophy (Johnston & Dunne, 1991). Scientific and mathematical inquiry, portrayed as logical, systematic and absolute, is presented as the most legitimate way of knowing reality and establishing truth. The powerful social and political consequences of elevating certain ways of knowing over others is hidden by the positivist claim that this knowledge describes reality and is part of a 'natural order' (Harding, 1986).

The traditional epistemology of science underpinned by the positivist paradigm has been critiqued vigorously over the last four decades (Phillips, 1987). Philosophers such as Popper (1968), Kuhn (1970) and Lakatos (1976) have rejected the 'institutionalised rationality' of positivist science by exposing as unsustainable central assumptions concerning the objectivity of truth, the role of evidence and the invariance of meaning. This critique has led to a recognition that knowledge is a human construction and that the processes by which it is legitimated are political, whether that knowledge is scientific, aesthetic or moral. Such a shift characterises the move towards Habermas' second interest, the practical.

The practical interest centres on developing understanding through interaction with the environment. Knowledge is recognised as socially constructed through interpretation and as verified through consensus. The acknowledgment of subjectivity highlights the moral dimension of knowledge production (Grundy, 1987). In the technical interest, the moral dimension is hidden by reference to objectivity, whereas the practical interest is linked by Habermas with hermeneutics and interpretive research.

Perhaps the most important issue which has characterised educational research in recent years has been debate about methodology, especially the apparent dichotomy that has been constructed in terms of quantitative versus qualitative methods. Applying Habermas' framework to these research positions raises challenging questions about what such research can tell us. Arguably the fundamental epistemological questions about the construction and legitimation of knowledge, in which all research is implicated directly, have been suppressed successfully by this focus on 'opposing' methods.

Generally, in quantitative research, data are collected to describe some 'reality', usually by researchers who are positioned outside the research arena. The task of the researcher is to quantify the variables which are the focus of the research, employing statistics to describe this 'reality'. Often there is no contact between the researcher and the researched and, in cases where there is interaction, the relationship might be described as one of 'inspection'. In educational research, teachers and students are often the objects of the research, being represented by, and manipulated as, quantifiable variables.

The quantitative approach is consistent with the claims to neutrality and objectivity associated with Habermas' technical interest. While quantitative research methods are not associated exclusively with the positivist paradigm, they are characteristic of it. As Stanley and Wise (1983, p. 173) point out:

The positivist research style, and the belief in one social reality, appears to be useful because it seemingly enables us to find things out. But what positivist research finds out is what the researcher already knows, in terms of the knowledge already existing within particular disciplines; and it might be seen as an efficient means of 'proving' to others what the researcher already knows is really 'true'.

Qualitative research, on the other hand, is a broad descriptor that includes a variety of methods, including participant observation, interviews and case studies. It often is referred to as naturalistic or interpretive research. In education, qualitative methods often are used when the purpose of the research is to gain some understanding of the meanings constructed by students and/or teachers and the intent of their actions. The researcher observes the actors in the arena (for example, the home, the classroom or the playground), taking account of the social interactions within it, and collecting data on which to base judgements and recommendations. The influence of the researcher is an important and contentious issue in qualitative research. Some researchers claim to neutralise their influence upon the observations by using methodological techniques such as 'triangulation'. Others, such as ethnographers, declare their position, but rarely see it as contestable because the researcher's interpretations are given greater epistemological status than those of other participants in the research arena.

Such research often is associated with Habermas' practical interest and highlights another dimension of the contradictions within that position. Assumptions about the social construction of knowledge clearly influence this kind of research but are limited within the research arena. Questions about whose accounts and whose interpretations constitute and organise knowledge are not raised explicitly. Rather they are submerged in the taken-for-granted organisation of education. Stanley and Wise (1983, p. 160) comment:

'Naturalism' is essentially 'dishonest', in the sense that it too denies the involvement, the contaminating and disturbing presence, of the researcher. Here too, just as in conventionally positivist research, we necessarily look at events *through* the researcher; but, in spite of this, such research is presented to us in such a way as to deny this, to suggest that what we have instead is 'truth'.

Although teachers' and students' accounts and interactions can be the basis for the interpretations made in the research, in the final analysis it is the researchers' accounts that are considered the most legitimate. The issue here is that the basis for that legitimation is not raised. Carr and Kemmis (1986, p. 216) criticise the normalisation of the hierarchical nature of the researcher/researched relationship:

The separation of theory and practice endemic to positivist and interpretivist views of research is now institutionalised in a division of labour between theorists and practitioners.

The academic division of labour produces and reproduces a differentiated epistemological status between the 'expert' researcher and the research participants. The analytical status of the researcher's account is juxtaposed with the descriptive status of the research participants' account. In addition, access to publication, which is facilitated through the higher (sic) education institutions, is a significant part of the process through which knowledge is constructed and legitimated.

THE EMANCIPATORY INTEREST

To this point in the chapter, through our development of the technical and practical interests, and the consideration of research methods in education, the processes of knowledge production and legitimation have been problematised. Such a critique represents a move towards Habermas' third category, the emancipatory interest, which is the most difficult of all three categories to conceptualise. In many ways, the emancipatory interest can be understood as a critique of the other two positions, a critique which centres on the absence of an explicit recognition of the politics of knowledge constitution within the technical and practical positions. As we have explored in the previous section, the claims within the technical interest to value-neutrality and objectivity, and the reference to 'natural laws', deny the social construction of knowledge and therefore hide its political nature. Within the practical interest, knowledge is regarded as socially constructed and deference is given to the views of 'experts' in the validation process. The claims to consensus, which, in fact, represent the views of the 'experts', serve to distance the practical interest from political considerations.

In the emancipatory interest, knowledge is regarded as socially constructed but the politics of its construction and legitimation are a central and continual concern. The implications of the emancipatory interest for educational research emanate from the centrality of critique, with an overt focus on power relations. By addressing explicitly the questions about knowledge production and legitimation, emancipatory research explores the social structures which work to maintain and reproduce the interests of the power holders. These power relations are operationalised at the level of specific contexts. Through critical analysis, the social and political contexts which circumscribe the production of knowledge are made explicit and recognised for the constitutive role which they play in the production and validation of that knowledge. Thus, in all educational research, the central concepts and categories around which the research questions are formed must be interrogated. Common sense assumptions cannot be left unquestioned, nor can they be considered in isolation from the broader issues of educational research, such as those relating to methodology.

In contrast to the other two interests, the outcomes of emancipatory research cannot be prescribed. The goals of technical and practical research are product-oriented in that the outcomes eventually are disassociated from the research process. The goals of emancipatory research, however, are realised in the processes of analysis of the power relations that circumscribe knowledge production (Stanley, 1990). In the technical and the practical interests, the privilege accorded to the researchers' accounts is not questioned. By contrast, the emancipatory interest explicitly addresses how researcher privilege constructs the contexts of the research. Further, methodological questions in the technical and practical interests are concerned with ensuring the validity and reliability of results, that is, that the results represent the 'truth' as closely as possible — whether it is perceived as absolute or socially constructed. In the emancipatory interest, the methods of data collection are not assessed according to the adequacy of their description of reality, but according to how they support and critique the social constructs of the specific research contexts and the broader social theory.

In order to explore these questions more fully, we now consider research into gender and mathematics/science education, providing a brief overview of the typical types of research in the area, and considering the conceptualisation of gender within these strands.

PROBLEMATISING GENDER RESEARCH

In surveying the research into gender and mathematics/science education over the last 20 years, certain trends become apparent. For example, one strand of research is concerned centrally with finding and documenting differences. Using quantitative methods, this research focusses upon differences in achievement or participation. Rarely do these studies seek explanations as to why or how such differences come about. Generally, they are used to monitor a particular situation over time, and to identify and document any shifts. There are many examples of such studies, including those for which statistics are gathered to compare the enrolments and achievements of females and males in science and mathematics (Dekkers *et al.*, 1986; Doron, 1991; Jones, 1991; Keeves, 1992; Lock, 1992; Megaw, 1991; Reilly & Morton, 1991).

Another strand of research seeks explanations for the apparent differences between the sexes in performance in mathematics and science by appealing to biology, as in studies into spatial ability (Smith, 1964). These investigations assume that there are differences in aptitudes between boys and girls which have innate causes, and attempt to isolate the nature of these differences (Gray, 1981; Sherman, 1983).

In Habermas' terms, both of these strands are located within technical interests. The research methodology assumes the authenticity of the categories used, the reliability of scientific methods to describe accurately the differences and, hence, the neutrality of the knowledge produced. Research that falls within the technical interests is predicated on the assumption that the environment can be understood and described, and is predictable and ultimately controllable. What counts as knowledge thus is not open to question – it is revealed through observation and can be verified through experimentation. In this research, the gender categories are considered to be discrete and natural and their differences are assumed to pre-date the situation being described in the research. They are regarded as part of the 'natural order', outside social influence and outside the influence of the researcher, whose job is to describe accurately the research site in terms of these categories. The purpose of the research is to reveal these differences, which are the logical consequences of the existence of the gender categories, by the use of sufficiently objective research methods. Thus the implication is that the differences described reflect innate characteristics of the members of each category.

Seeking social explanations for gender differences constitutes a third strand of research. Studies attempt to link students' perceptions and attitudes to a masculine image of science and mathematics, examine the effects of parental or teacher attitudes and practices, and investigate the influence of sex-stereotyping in the home and at school (Ethington, 1992; Fennema, 1993; Kelly, 1987; Leder, 1974, 1976, 1982; Murphy, 1991; Sjøberg & Imsen, 1988). Attempts to change the existing situation have given rise to strategies which have been devised, implemented and, in some cases, evaluated (see, for example, Kelly, 1987). Some strategies concentrated on changing the girls by attempting to change some of their learning characteristics to match, more closely, those of the boys. Another set of strategies has concentrated on altering the curriculum to make it more gender-inclusive (Barnes, 1991; Gianello, 1988). Yet others suggest that changes in the early socialisation patterns, learning environments or the school organisation can have an impact (Barnes *et al.*, 1984; Burton, 1986, 1990; Lewis & Davies, 1988).

We locate this strand in Habermas' practical interest. The acknowledgment of the social construction of gender emanates from a moral position of concern for equal opportunity. Much of the research is qualitative in nature, drawing upon interpretive, naturalistic and hermeneutic methods. While these research methodologies recognise the importance of context and setting in the mediation of meaning, they assume also that the product of the research will describe adequately the research site in terms of its social processes and the understandings constructed. The aim of such research is to describe, as accurately as possible, the 'reality' of mathematics and science learning environments for girls and boys. As in the former two strands, this research acknowledges a biological basis for gender categories, but places emphasis upon the role of social practices in constituting the differences. It is both biology and interactions in the social environment which produce the gendered individual. Interpretive research in this field describes the site by looking for differences in the experiences and interactions of females and males. The assumption is that the differences observed and described characterise the members of each category and, in effect, those experiences and interactions assume the status of essential gender characteristics.

Statements arising from research within the technical and practical interests are taken to represent the 'truth' about girls, boys, mathematics and science. These include statements such as 'girls prefer collaborative learning environments', 'girls need opportunities to use their language skills', 'girls prefer to share and support each other in tackling problems', 'boys perform better in competitive situations than girls', and 'girls need encouragement to build their confidence and self esteem'. A change in the situation, then, requires either girls to have experiences that compensate for their deficiencies (for example, providing play with spatial toys, encouraging assertive behaviour, helping them to make better choices), or for the school learning environment to be altered to compensate for the learning styles of girls (for example, group work, writing in mathematics, emphasis on the human applications of science). In seeking to affirm the experiences of girls in school mathematics and science, gendered oppositions are noted (or, in our terms, are constructed) and these provide the explanations for interventions. Social constructs, such as collaboration and competition, dependence and independence, compliance and aggression, are inscribed upon the original biological divisions and are assumed to be characteristics of individuals in each category. Thus the social and biological are conflated in the production of 'essential' characteristics.

What must be recognised here is that the oppositions that are constructed, within both the research and the interventions which are developed from it, are constitutive of gender. They produce and reproduce the categories that they are assuming to describe. Ironically, in this production, the relationship that the research is seeking to challenge – the dominance of the masculine over the feminine – is reproduced through these oppositions. Girls, as collaborative (in terms of preferred work mode), dependent (on each other and the teacher) and compliant (to the demands of the classroom, the curriculum and the researcher) are differentiated from boys who are competitive, independent and aggressive. But, in terms of learning in mathematics and science, it is competition, independence and aggression that are valued. The masculine side of the opposition is defined positively and treated as the norm. Success in mathematics and science demands such attributes. The feminine side of the opposition is positioned negatively in relation to this norm and to school mathematics/science (see, for example, Walkerdine, 1984, 1989).

Grosz (1990) warns of the limitations of the views of gender that we have described as typifying the three strands of research:

(I)n claiming that women's current social roles and positions are the effects of their essence, nature, biology, or universal social position, these theories are guilty of rendering such roles and positions unalterable, necessary, and thus of providing them with a powerful political justification . . . they are necessarily ahistorical; they confuse social relations with fixed attributes; they see fixed attributes as inherent limitations to social change. (Grosz, 1990, p. 335)

Such positions are unlikely to produce significant change, and worse, can be used to justify a lack of change. The essentialisms that Grosz describes stem from inadequate conceptions of social relations. Almost all of the research that we have described here presumes the autonomy of the actions of individual human subjects in considering or describing the research context. The research that we have described as technical has no conception of the social. The focus of the research is upon groups of individuals who share common characteristics, in this case their sex, and on describing how those groups perform under certain conditions. The research that we have described as practical considers the influence of the social in terms of the interaction of each individual with the learning environment. Constructs used to describe social interaction become defining attributes of gender categories and change is only possible if the individuals, in this case girls, adjust their attributes or their choices (Willis, 1995).

From the emancipatory position, the emphasis on social relations rather than a delineation of fixed attributes is highly significant. It transforms gender research from descriptions based on static 'common sense' categories to a consideration of the construction of these categories. This dynamic is a manifestation of power relationships which are realised through the construction of differences. Gender, then, must be conceptualised in ways in which the construction of difference within the contingencies of a specific context are made explicit. In considering it as a dualistic and hierarchical relation, produced and reproduced by social practices, it cannot be conceived of as pre-given, either by biological or social means.

CONCLUSION

That there is 'an issue' about gender and school mathematics/science is now accepted widely, although conceptions of what the problem is have changed (Willis, 1989, 1995). These conceptions have ranged from the position that gender differences in performance and participation are 'only natural', to a view that the psychological explanations of innate capacities is required, to considerations of the impact of early socialisation or the conditions of schooling on girls' participation in mathematics and science. In this chapter, we have considered what it is that each of these strands of research has taken as the object of its research, that is, what it is that constitutes gender. We have done so using Habermas' framework of knowledge-constitutive interests, in which his third interest, the emancipatory, provides a critical position which foregrounds explicitly the political nature of knowledge constitution.

Assuming a 'critical' position is not uncommon in social research today. In educational research, the term has been applied broadly to philosophical and methodological positions which emphasise the political nature of schooling, usually addressing social inequities constructed around gender, class and racial differences which are perpetuated through schooling. Stressing that no knowledge is politically neutral, that all institutions are ideologically bound, emancipatory research aims to problematise the institutional foundations of knowledge and to make explicit the investments in power which constitute those foundations.

Our consideration of the role of research in the construction of gender categories is one attempt to problematise the fundamental assumptions of the research process. For research into gender and mathematics/science education, an emancipatory perspective demands that the categories used to delimit the construct of 'gender' are addressed explicitly in terms of their construction. An acknowledgment of social processes in the construction of gender is not sufficient. What is required is an engagement with the dynamics of gender construction, that is, with the production and reproduction of this dualistic relation in and through social practices. Research from the practical and technical interests is predicated upon fixed categories as if these exist as fixed characteristics prior to the research process. From our critical position, the focus shifts to the production of these categories, both in the social practices that constitute the empirical site and in the processes of framing and undertaking the research.

For policy makers and practitioners, as for researchers, a shift in the conceptualisation of gender demands an acknowledgement and interrogation of how the practices in which they are engaged are implicated in the production and reproduction of gender difference. Such a focus does not imply that the research processes and tools that have served us in the past, or the research findings applied to policy making and teaching, are no longer useful. However, their use is problematised. Not only are the foundations of the traditional disciplines an explicit focus of attention, but so too are the assumptions, understandings and taken-for-granted practices which underpin research processes, as well as those which define policy-making and teaching. Thus, with respect to gender and mathematics/science education, we need to engage in a continual process which acknowledges, and seeks to understand, how gender is constructed in the research process itself, and in the appropriation of the research procedures and findings into practice. This position of critique is not an easy one to sustain — not the least because it demands an undermining of the technical and practical foundations on which our practices are based. Nevertheless, we contend that, by problematising the research processes, and making explicit some of the assumptions that underpin our practices, fundamental questions about the interaction between gender and power in our society will be raised. Thus, the privilege that access to mathematical and scientific knowledge bestows on its 'chosen few', and the 'gate-keeping' role that such knowledge plays in relation to further education and future careers, need to be understood in relation to the production and reproduction of hierarchical gender (and class and race) positions. Critical questions are what constitutes mathematics and science, what counts as valued knowledge, and how things came to be this way and how they are sustained. Such questions foreground relations between power and knowledge that exemplify the emancipatory position.

¹Education Department of Western Australia, Perth, Australia;
²The University of Birmingham, UK

REFERENCES

Barnes, M. (1991). *Investigating change: An introduction to calculus for Australian schools*, Melbourne, Curriculum Corporation.
Barnes, M., Plaister, R. & Thomas, A. (1984). *Girls count in mathematics and science: A handbook for teachers*, Sydney, Mathematical Association of New South Wales.
Burton, L. (ed.) (1986). *Girls into maths can go*, Holt, London, Rinehart and Winston.
Burton, L. (ed.) (1990). *Gender and mathematics. An international perspective*, London, Cassell.
Carr, W. & Kemmis, S. (1986). *Becoming critical: Education, knowledge and action research*, London, Falmer Press.
Dekkers, J., de Laeter, J.R. & Malone, J.A. (1986). *Upper secondary school science and mathematics enrolment patterns in Australia, 1970–1985*, Perth, Curtin University of Technology.
Doron, R. (1991). 'Gender similarities and dissimilarities in prediction of academic achievements by psychometric tests among Israeli practical engineers', in L.J. Rennie, L.H. Parker and G.M. Hildebrand (eds.), *Action for equity: The second decade. Contributions to the Sixth International GASAT Conference*, Perth, National Key Centre for Teaching and Research in School Science and Mathematics, Curtin University of Technology, 544–552.
Ethington, C.A. (1992). 'Gender differences in a psychological model of mathematics achievement', *Journal for Research in Mathematics Education* 23(2), 166–181.
Fennema, E. (1993). 'Teachers' beliefs and gender differences in mathematics', in E. Fennema and G. Leder (eds.), *Mathematics and gender*, Brisbane, University of Queensland Press, 169–187.
Gianello, L. (ed.) (1988). *Getting into gear, gender inclusive teaching strategies in science*, Canberra, Curriculum Development Centre.

Gray, J.A. (1981). 'A biological basis for the sex differences in achievement in science', in A. Kelly (ed.), *The missing half: Girls and science education*, Manchester, Manchester University Press, 43–58.

Grosz, E. (1990). 'Conclusion: A note on essentialism and difference', in S. Gunew (ed.), *Feminist knowledge, critique and construct*, London, Routledge, 332–344.

Grundy, S. (1987). *Curriculum: Product or praxis*, London, Falmer Press.

Habermas, J. (1972). *Knowledge and human interests*, second edition, London, Heinemann.

Harding, S. (1986). *The science question in feminism*, Milton Keynes, Open University Press.

Johnston, J. & Dunne, M. (1991, April). *Gender, mathematics and science: Evading the issues or confronting new questions*, Paper presented at the annual meeting of the American Educational Research Association, Chicago, IL.

Jones, M.G. (1991). 'Competitive science: Gender differences in the physical and biological sciences', in L.J. Rennie, L.H. Parker and G.M. Hildebrand (eds.), *Action for equity: The second decade. Contributions to the Sixth International GASAT Conference*, Perth, National Key Centre for Teaching and Research in School Science and Mathematics, Curtin University of Technology, 261–269.

Keeves, J. (ed.) (1992). *The IEA study of science III: Changes in science education and achievement: 1970 to 1984*, Oxford, Pergamon Press.

Kelly, A. (1987). 'Why girls don't do science', in A. Kelly (ed.), *Science for girls?*, Milton Keynes, Open University Press, 12–17.

Kuhn, T.S. (1970). *The Structure of scientific revolutions*, Chicago, IL, University of Chicago Press.

Lakatos, I. (1976). *Proofs and refutations*, Cambridge, Cambridge University Press.

Leder, G.C. (1974). 'Sex differences in mathematics: Problem appeal as a function of problem context', *Journal for Educational Research* (67), 351–353.

Leder, G.C. (1976). 'Contextual setting and mathematical performance', *Australian Mathematics Teacher* 32(4), 119–127; 32(5), 165–173.

Leder, G.C. (1982). 'Mathematics achievement and fear of success', *Journal for Research in Mathematics Education* 1(2), 124–135.

Lewis, S. & Davies, A. (1988). *Girls and maths and science teaching: Professional development manual*, Canberra, Curriculum Development Centre.

Lock, R. (1992). 'Gender and practical skill performance in science', *Journal of Research in Science Teaching* (29), 227–241.

Megaw, W.J. (1991). 'Gender distribution in the world's physics departments', in L.J. Rennie, L.H. Parker and G.M. Hildebrand (eds.), *Action for equity: The second decade. Contributions to the Sixth International GASAT Conference*, Perth, National Key Centre for Teaching and Research in School Science and Mathematics, Curtin University of Technology, 604–612.

Murphy, P. (1991). 'Gender differences in pupils' reactions to practical work', in B. Woolnough (ed.), *Practical science*, Milton Keynes, Open University Press, 112–122.

Phillips, D.C. (1987). *Philosophy, science and social inquiry: Contemporary methodological controversies in social science and related applied fields of research*, Oxford, Pergamon.

Popper, K.R. (1968). *The logic of scientific discovery*, London, Hutchinson.

Reilly, B. & Morton, M. (1991). 'Performance in a nationwide mathematics examination at tertiary entrance level', in L.J. Rennie, L.H. Parker and G.M. Hildebrand (eds.), *Action for equity: The second decade. Contributions to the Sixth International GASAT Conference*, Perth, National Key Centre for Teaching and Research in School Science and Mathematics, Curtin University of Technology, 301–309.

Sherman, J. (1983). 'Girls talk about mathematics and their future: A partial replication', *Psychology of Women Quarterly* (7), 338–342.

Sjøberg, S. & Imsen, G. (1988). 'Gender and science education I', in P. Fensham (ed.), *Development and dilemmas in science education*, London, Falmer Press, 218–248.

Smith, I. (1964). *Spatial ability: Its education and social significance*, San Diego, CA, Robert P. Knapp.

Stanley, L. (1990). *Feminist praxis*, London, Routledge.

Stanley, L. & Wise, S. (1983). *Breaking out: Feminist consciousness and feminist research*, London, Routledge and Kegan Paul.

Walkerdine, V. (1984). 'Developmental psychology and the child centred pedagogy: The insertion of Piaget into early education', in J. Henriques, W. Hollway, C. Urwin, C. Venn and V. Walkerdine, *Changing the subject: Psychology, social regulation and subjectivity*, London, Methuen, 153–202.

Walkerdine, V. & The Girls and Mathematics Unit (1989). *Counting girls out*, London, Virago Press.

Willis, S. (1989). *'Real girls don't do maths': Gender and the construction of privilege*, Geelong, Australia, Deakin University Press.

Willis, S. (1995). 'Gender reform through school mathematics', in P. Rogers and G. Kaiser (eds.), *Equity in mathematics education: Influences of feminism and culture*, London, Falmer Press, 186–200.

SECTION II

THE REALITY OF SCHOOLS, CLASSROOMS, CURRICULUM AND ASSESSMENT

TERRY EVANS

6. UNDER COVER OF NIGHT:
(RE)GENDERING MATHEMATICS AND SCIENCE EDUCATION

THE GENDERED WORLD

There are many stories about science. The ones which children read or have read to them typically present a benign view of a world of bespectacled, potty professors who invent magic potions, cures and remedies to help children and other people or animals out of some calamity. Sometimes they construct rickety rockets or shiny spaceships to transport children and themselves to other galaxies in pursuit of another adventure. These are the children's bedtime stories which take them off to their slumbering dream worlds for yet another night. What goes on during the night? How do these early stories interplay with the children's emerging understandings of the worlds in which they live? What foundations do they lay for the future? How do the girls and boys see themselves in their dreams?

What goes on during the night? A cement works nearby pushes its yellow-grey clouds into the air, a chemical works adds its noxious gases, and a brown coal power station provides the twist of tar to this nocturnal cocktail. All this occurs at a time and within the limits prescribed by the legislators, on the advice of scientists. Across the world, at night, a railway train carries another flask of radioactive materials from the naval docks, past sleeping children, to a nuclear power station miles away. In the heart of the city, a laboratory technician monitors the instruments and records the data through yet another long night of experimentation. Nearby in the hospital, a nurse administers the four-hourly medications to the young children who have recently undergone surgery; some of their parents are dozing nearby. At the airport, two engineers re-program the computers for the air traffic control systems before the night time curfew is lifted and the first scheduled flight is due.

The children's dream worlds and the adults' 'science' worlds are not just separations of fiction from fact. Indeed the fiction contains facts – there are science professors (some even could be potty!), they are usually men, and there are rockets and spaceships. And the facts contain more than a few fictions – the 'objective' nature of science, science equals 'progress', etc. Between the nights of the children and the nights of science, there spans a flow of socialisation which sees some of the child dreamers take their places as the plant operators, engineers, laboratory technicians or weapons officers of the future realities. Yet, when we critically analyse those future realities, we can see that the people who occupy most of the places in the industrial, military and research worlds of science are men; only the laboratory technician is more likely to be a woman. In the medical world of science, the nurse is more likely to be a woman, and the surgeon a man. Yet we know that these gender

67

L.H. Parker et al. (eds.) Gender, Science and Mathematics, 67–76.
© 1996 Kluwer Academic Publishers. Printed in the Netherlands.

divisions are neither absolute – men are nurses, women are engineers – nor universal.

Simone de Beauvoir has noted how, in the former Soviet Union, most doctors were women 'because medical treatment is free' (Schwarzer, 1984, p. 30) and yet 'the really important jobs, like those in science and engineering, are much less accessible to them' (p. 32). She points to how 'there is the same scandalous situation in Russia as there is elsewhere . . . housework and looking after children are exclusively female preserves too' (p. 32). De Beauvoir notes how this is reflected in the stories from Soviet novelists: 'This comes over strikingly in Solzhenitsyn's (1971) *Cancer Ward*. There is a woman in the hospital who is very senior, a very important member of the medical profession; after doing her rounds at the hospital, she rushes off home to cook for her husband and children, and to do the washing-up' (p. 32).

We can contrast this with the life of a man in a similar senior position: he would have his meals cooked, home cleaned, children looked after and clothes laundered. Men who pursue careers in science and technology, or as science teachers, are no different from other men. When they marry (or form similar relationships with a woman), men usually exchange one domestic labourer for another: mother for wife or partner. David Suzuki, for example, began his adult life as a typical research scientist. He immersed himself in scientific study and research with the goal of becoming a leading geneticist in a university or research institute. He married while he was pursuing his PhD and his wife helped support their income as a laboratory technician. In his autobiography, entitled *Metamorphosis: Stages in a Life* (1988), Suzuki explains how his family life worked for him during his early career in the 1960s. After the birth of their first child, his wife, Joane, returned to work and between them they looked after their daughter. As a PhD student, he adjusted his schedule so that he worked late at night and looked after their child, Tami, during the day; Joane returned from her work during the day to look after their child at night. However, as Suzuki's career in science continued into his post-doctoral year another child, Troy, was born:

I was well into my research and thinking about getting a permanent position. Tami was continually surprising us with each new step in her development, and as he grew Troy was charting now familiar territory . . . (H)e was my father's first male grandchild, and therefore, in the continuing pattern of Japanese thinking, very special. Troy was a wonderful child, good natured and understanding; it's ironic that had he been a problem child I might have given him more of my time.

Joane was a marvellous mother and a good wife. During those times, she bore my long nights and weekends at the lab and never complained. I called the shots – deciding when and where we'd take a holiday, when I'd come home for dinner, whom we'd have over. And all the while I couldn't share with my wife the excitement of what was going on in the lab. I'm sure that when I said 'I'm going to the lab' to Tami and Troy, they imagined me disappearing into some dark and mysterious hole. (Suzuki, 1988, pp. 173–174)

After this period, Suzuki obtained his first academic position at the University of Alberta:

I had to set up a new lab, start my experiments, teach a new course, and apply for a research grant. If (the postdoctoral year) had taken a lot of time it was nothing compared to (the new university faculty position). I hurled myself into my teaching and found my social life revolving around my students.

This drew me further away from my home life. I don't even know now what Joane did with the children, especially during the long cold winter in Edmonton. When I came home for dinner, the children were there for me to play with, bathe, read to and tuck in bed. The family seemed to be there waiting for my whim and inclination. (Suzuki, 1988, p. 174)

Suzuki separated from his wife at around the time their third child was born. He later married 'a committed feminist' (p. 288) and had another child in the 1980s, when the expectations of him were somewhat different both institutionally (for example, he was now expected to be at the birth of his children, whereas for his previous children he was expected not to be present) and personally (through his encounter with feminism). This led Suzuki into a period of reflection on his first phase of parenthood:

Only now that I have a second family with small children, do I realise the enormity of what Joane did in raising those children alone. I thought I was a dutiful father putting in my time with the children. But children's needs cannot be accommodated to a parent's schedule . . . I know the slogan 'it's *quality time not quantity* time that matters' and I think it's baloney. I failed my first family and my failure was not being there in those moments when noses need wiping, tears brushed away. . . . I was too selfish and self-centred, and the children paid the price. (Suzuki, 1988, p. 189)

What such vastly different people as de Beauvoir (and, unwittingly no doubt, Solzhenitsyn) and Suzuki are telling us is that typically the people who pursue careers in science (in the broadest sense) are fettered to the gender-structured nature of those careers and of the broader social life. What goes on during the night? The budding scientist Suzuki goes the lab or, when he is in his parental role, bathes the children and reads them a story. Yevgenia Ustinovna, the senior surgeon in Solzhenitsyn's *Cancer Ward* (1971), spends her nights doing the domestic chores. In this sense, science careers are similar to many other careers – including, importantly for this discussion, teaching – in that those men or women who 'dedicate' themselves to their work are seen as virtuous and good for society. Crucially the politicians and other key policy makers, such as senior bureaucrats, business and union leaders, all share similar values.

WHAT CHANCE IS THERE FOR CHANGE?

The chances for change are not very good. Unless one is prepared to be very patient and those who are striving for change are prepared to be very diligent, change is going to be elusive. But there are sufficient examples, in the biographies of women from different generations, which show that the shackles of gender have been loosened a little; and they have been loosened a little for men, too. It is important to remember that various social processes operate to maintain the status quo and to reproduce the social structures of society, such as those concerning gender. These social processes are 'cradle to grave' and, as was shown at the beginning of this chapter, pervade not only the daylight hours of public life but also the night times of our private worlds. Schooling is one of the important ways in which gender structures are reproduced *and* lived out in everyday life. I have argued elsewhere that the adult members of school communities form a *'living curriculum'* for the children to engage in and learn (Evans, 1989) and that teachers' perceptions of gender shapes their work as teachers (Evans, 1982). Schools are a social context in which (amongst other things) various forms of gender-structured personal biographies come together to construct and re-construct gender and other social divisions (Evans, 1988).

GENDER AND THE TEACHING AND LEARNING OF SCIENCE

In focussing on the relationships between gender and the teaching of science in schools, it is worth considering the gendered nature of science itself, because it is this which frames or prefigures the science curricula. Several writers have been able to chart the difficulties which befall women who seek to pursue careers in science (Gornick, 1983; Kahle, 1985; Keller, 1985) and others have noted the ways in which gender relations shape the everyday life of an Australian scientific community to produce a range of problems, contradictions and barriers for women – and, in different sense, for men (Charlesworth *et al.*, 1989). Despite its esteem as a 'rational and objective' area of human endeavour, science has at least its fair share of the patriarchal trappings of society.

Science, from Newton onwards, has been seen by some scholars as largely in the service of capitalism and the state's material and military production demands (Mulkay, 1979). More recent feminist critiques have demonstrated the basic masculine nature of contemporary science and shown, like Keller (1985), that this nature is as much a part of the language and culture of science as it is embedded in the organisation and practices of scientists themselves.

Is it realistic, fair and reasonable to endeavour to achieve gender equity in the participation of students in mathematics and sciences at school and in scientific careers if science itself, both as a body of knowledge and skills and as an array of careers and practices, is profoundly gender structured and inequitable? The question arises as to what science and whose science is to be taught at school? It might be partially successful, for example, to encourage and support girls into science in the same way as this has occurred with encouraging and supporting women into 'non-traditional' occupations. One might hope that successive generations of girls in science or women in 'non-traditional' occupations will bring about a change to the nature of science or those occupations. But why should girls and women in these circumstances be expected to take on such pioneering work? Why should they be expected to shoulder the burdens of previous centuries of traditional gender structures and encounter contemporary sexism as well? Reasonably, they might be expected to make their contribution, but being a pioneer seems to be impractical from a schooling viewpoint and to ignore the nature of gender in science and society. Men should recognise and shoulder their responsibilities in these matters.

The point about the gendered nature of science, schools and society is that it is dominated by men, not simply through occupying most of the powerful positions, but through the culture itself sustaining masculine superiority. This culture is formed by men and women and is perpetuated by (or can be resisted by) both men and women. If we accept that men dominate through the actions of both men and women, then it seems reasonable to argue that: in order to understand the nature of gender in society and the domination of men in most forms of social life, one needs to take a gender-critical perspective of both men and women and the interrelations between them (Evans, 1988); and, in order to bring about any enduring gender equity, men and women need, separately and collectively, to reach a gender-critical understanding and make the changes to their lives accordingly. To suggest that, in the case of mathematics and science teaching, the attention should be focussed on girls is not only missing the point about gender in science, school and society, but it also blames the victims. Keller provides us with a useful insight here:

> The widespread assumption that a study of gender and science could only be a study of women still amazes me: if women are made rather than born, then surely the same is true of men. It is also true of science. . . . My subject is not women per se, or even women and science: it is the making of men, women and science, or more precisely, how the making of men and women has affected the making of science. (Keller, 1985, pp. 3–4)

Keller's argument holds good for other areas of gender study, especially education. It is assumed that the study of gender and education is the study of women in education, by women. What occurs in this line of thinking is a conflation of women's studies and gender studies. This is not to deny the important theoretical, methodological links and overlaps between the two areas of study, but any involvement of men in women's studies poses many problems which need to be addressed on women's terms. Conversely, the lack of involvement of men in gender studies poses problems that women cannot address effectively. Without men, gender studies will be a partial and distorted discourse, and actions directed at gaining gender equity in mathematics and science will be unsuccessful. This can be seen, as Behringer records, in the biographies of women in science: 'Historically, many women who either sought to study science or to work in science were aided and abetted in their struggles by concerned men' (1985, p. 40). It is arguable that any girl who is to make her way through science at school and university and through a 'successful' science career will be unable to do so unless some of the ('concerned') men she encounters 'aid and abet' her. An important task, therefore, is to raise the concerns of men in general – and men in science-based work, including science teachers in particular – so that 'aiding and abetting', in the conspiratorial sense, is no longer the process. 'Aiding and abetting' becomes redundant as the 'unconcerned' men lose their dominance and grip to the emerging majority of men and women who have a gender equitable science in their minds.

A PROBLEMATIC RATIONALE

The problem is that the increase in the gender concerns of men at present seems less to do with gender equity and more to do with exploiting the talents and skills of women in the service of the economy and state. This gives the illusion of gender equity, and it could increase the proportions of women in mathematics and science careers, but their purpose is more related to maintaining traditional forms of class and gender structures. The Review Panel for the *Discipline Review of Teacher Education in Mathematics and Science* made the following point:

The emerging rationale is one of system needs. Science and technology need to have significantly more people (and brighter people), and girls are an underutilised source. Or science and technology need to have significantly more people with different ways of thinking . . . including thinking which is more person-oriented, more cooperative than competitive in orientation, more concerned for the environment, for example, and . . . these ways of thinking are more common among women than men. (Speedy *et al.*, 1989, p. 140)

Although the Review Panel did not adopt these arguments overtly itself, it allows the credence of the 'rationale' to be sustained in its own discussion. These arguments are not new, although they are coloured by the economic, political and labour market conditions of the times. For example, such arguments – and their relationship to women's childcare responsibilities, girls' choices of subjects at school, etc. – are to be found a decade earlier in the reports commissioned by the Australian Government into *Education, Training and Employment* (Williams, 1979) and *Technological Change in Australia* (Myers, 1980).

The popularity of this view about women and science has been engendered through the mass media. The Australian science broadcaster, Robin Williams, gives as one of his reasons why girls must do science that 'this country will soon be desperate for scientists. . . . One reason it's so important to have girls (in science) is that they bring a new spirit of enquiry to a team. . . . Girls tend to have a wider framework of interest, not because they are built differently, but because our world points them towards more culture. This helps enormously in science' (1990, p. 15). One cannot fault Williams's intentions. Indeed, his 'national interest' argument is likely to strike a chord with powerful men who see their interests as *being* the national interest. Therefore, such men, in their own interests, might resist women less, be less sexist or even seek out these girls with the 'new spirit of enquiry'.

Another way of interpreting Williams's argument is that we need more girls in science to compensate for deficiencies in the 'spirit of inquiry' and 'culture' of boys. This could be read as a good reason for improving the spirit and culture of boys, or even for removing some them from science. Hence, a practical consequence might be that boys need affirmative action programs to address their problems of spirit and culture. Should such an affirmative action program involve their female peers? Why should girls help boys with their problems, when boys create so many for them?

Williams's points about girls' ways of thinking also have some difficulties, although Williams is careful to avoid them himself. These difficulties centre on the popular opinion that boys and girls are different because of biology rather than culture. In this way, arguments about the differences between boys and girls, or in men's and women's thinking and behaviour, can be reduced conveniently to a matter of chromosomes, which schools and society at large can do little about. Such arguments need to be shattered with an array of examples for which men's and women's thinking and behaviour disturb this stereotypical 'chromosomal' notion.

One example will make the point. It was a woman who, apparently between a church service and her Sunday lunch, considered and then ordered the killing of 368 young men on 2 May 1982. She had been a research scientist, but it was in her capacity as the head of a government at war that she condoned and authorised the sending of these men to horrific deaths at sea at night. Whatever the merits or otherwise of Margaret Thatcher's decision to order the attack by the British submarine *Conqueror* on the Argentine cruiser *Belgrano* during the Falklands/Malvinas war, the point is that it was a woman's 'mind set' which informed and formed her decision, although, of course, it was made in the context of other people's thinking (predominantly men's) but no-one would doubt that Thatcher stands by her decision, and that she rests easily at night.

THE (RE)GENDERING TASK

I return now to the *Discipline Review of Teacher Education in Mathematics and Science*, at the time a highly influential document in Australian education. The Review contains two brief chapters which address 'access and equity' (six pages) and 'gender' (nine pages), where the majority of the discussion of gender takes place. Although arguably there are some contradictions within this discussion, some useful points are made. The Review Panel comments: 'In most of our meetings with senior academic staff in mathematics and science there was surprise and bewilderment when we asked about how gender issues were being taken into account in teaching' (p. 152). Clearly, apart from the exceptions noted in the report, most senior academic staff (and no doubt, a good number of junior ones too) in teacher education institutions could not recognise a gender issue in science or their own teaching if their careers depended on it! What needs to occur, at least in part, is that these academic staff in teacher education embrace some of the approaches that some of their colleagues and many more teachers in schools have been developing. The Review Panel comments that:

> . . . there is now quite a deal of evidence to suggest that the pedagogies that have been developed to encourage more girls in mathematics and science are also effective for many boys. The prospect that these inclusive pedagogies may enhance the pool of both girls and boys who succeed in, and enjoy, mathematics and science in school is one that should attract the attention of all those who are responsible for and involved in the teaching of these subject areas in higher education. (Speedy *et al.*, 1989, p. 151)

One would like to endorse the Review Panel's comments here, but such endorsement has to be tempered by the Review Panel's apparent lack of commitment to even its own brief observations on these matters. In its 12 or so pages of recommendations (pp. xxi–xxxiii and 205), not one specifically addresses gender, although there are some links from a few of these Recommendations (see, for example, 7.3) to gender-related points and to recommendations elsewhere in the Review's three volumes. What does go on during the night?

It is at moments like this that we can recognise the enormity of the task to redress gender issues in mathematics and science in schools, (i.e., to 'regender' them). We have seen that science itself embodies its own deep gender structures which shape the nature of the discipline and its practices. Now we can see that, not only is teacher education riddled with the same problems, but also that those charged with the responsibility for review and reform can be seen to equivocate. Teacher education really does have to address these matters because it has the opportunity to help teachers continue and extend the task that many of them already have started. This means developing mathematics and science education courses which are gender-critical and gender-inclusive and which regender the field accordingly; this is men's work too and this must be confronted. The issue of 'unconcerned' men who teach science and who, both consciously and subconsciously, denigrate girls' capacities and performance needs to be dealt with both in their teacher training and in their workplaces.

In many respects, an important area to concentrate upon in teacher education is teachers' continuing education and professional development. Some research into teachers who undertake distance education courses to improve their qualifications shows that teachers are keen to study in areas which have relevance to their professional practice and which 'update' their knowledge and skills (Evans & Nation, 1991, 1993). If some teachers are lacking in either their knowledge and skills in mathematics or in their understanding of the issues of gender in their teaching, then teacher professional development is an obvious way to address both of these matters. Again this points to the need for teacher education institutions to be an important focus and it means that the 'unconcerned' men in teacher education have to be confronted with their own prejudices and practices. It is a step backwards, in terms of gender equity in science and mathematics, if teachers who are keen to improve their professional qualifications and practices are faced with teacher education courses which reflect worse curricular and pedagogical practices than those of schools.

One assumes that all teacher education courses, not just those specifically concerned with gender, should address issues of a gender-critical kind. All teachers, especially those teaching teachers, need to encounter the sort of 'overnight' experience that Keller describes for herself:

I was deeply engaged . . . in my work as a mathematical biophysicist. I believed wholeheartedly in the laws of physics, and in their place at the apex of knowledge . . . (O)vernight, as it were – another kind of question took precedence, upsetting my entire intellectual hierarchy: How much of the nature of science is bound up with the idea of masculinity, and what would it mean for science if it were otherwise? A lifelong training had labeled that question patently absurd; but once I actually heard it, I could not, either as a woman or as a scientist, any longer avoid it. (Keller, 1985, p. 3)

That question needs to be projected into the thinking of everyone involved in science, especially those in mathematics and science teacher education who seem to have avoided it for so long. In this way, we might be able to assist, encourage and extend the moves that are already afoot in schools, rather than ignoring and undermining them with the sexist and patriarchal curricula and pedagogies of our universities. However, the broader issues concerning the ways in which domestic and parenting labour are gender-structured in society, especially in schools and science careers, also must be addressed. Keller argues:

The disengagement of our thinking about science from our notions of what is masculine could lead to a freeing of both from of the rigidities to which they have been bound, with profound ramifications for both. Taken together these two related tasks may contribute to a more equitable and fair society in terms of gender, for both women and men (1985, p. 92).

She rightly points out that her 'analysis rests on the significance of the gender of the primary parent, changing patterns of parenting could be of critical importance' (p. 93).

Both parenting and domestic labour need to be taken into account (Evans, 1988). In the present context, this is the case not just for scientists and teachers, but for teacher educators as well and, indeed, for us all. We need to recognise the interrelated nature of the public activities during the day, such as teaching mathematics and science at school, with the private activities that occur as night falls. It is important to peer into the dark spaces behind public activities, such as discipline review reports, curriculum workshops and science broadcasting, to see and change the underlying gender structures and processes which are pulling the next generation of career scientists into line.

Deakin University, Geelong, Australia

REFERENCES

Behringer, M.P. (1985). 'Women's role and status in the sciences: An historical perspective', in J.B. Kahle (ed.), *Women in science: A report from the field*, London, Falmer Press, 4–26.

Charlesworth, M., Farrall, L., Stokes, T. & Turnbull, D. (1989). *Life among the scientists: An anthropological study of an Australian scientific community*, Melbourne, Oxford University Press.

Evans, T.D. (1982). 'Being and becoming: Teachers' perceptions of sex-roles and actions toward their male and female pupils', *British Journal of Sociology of Education* (3), 127–143.

Evans, T.D. (1988). *A gender agenda*, Sydney, Allen and Unwin.

Evans, T.D. (1989). 'Living curricula: Adult gender relationships in school communities', in G. Leder and S.N. Sampson (eds.), *Education of girls: Practice and research*, Sydney, Allen and Unwin, 73–83.

Evans, T.D. & Nation, D.E. (1991). 'Teachers' professional development through distance education', in P. Hughes and P. McKenzie (eds.), *Teachers' professional development*, Melbourne, Australian Council for Educational Research, 114–128.

Evans, T.D. & Nation, D.E. (1993). 'Educating teachers at a distance in Australia: Some history, research results and recent trends', in H. Perraton (ed.), *Distance education for teacher training*, London, Routledge, 261–286.

Gornick, V. (1983). *Women in science*, New York, Simon and Schuster.

Kahle, J.B. (ed.) (1985). *Women in science: A report from the field*, London, Falmer Press.

Keller, E.F. (1985). *Reflections on gender and science*, New Haven, CT, Yale University Press.

Mulkay, M.J. (1979). *Science and the sociology of knowledge*, London, Allen and Unwin.

Myers, R. (1980). *Technological change in Australia: Report of the Committee of Inquiry into technological change in Australia*, Australian Government Publishing Service, Canberra, Australia.

Schwarzer, A. (1984). *Simone de Beauvoir today* (trans. M. Howarth), London, Chatto and Windus.

Solzhenitsyn, A. (1971). *Cancer ward*, Harmondsworth, Penguin.

Speedy, G., Annice, C., Fensham, P.J. & West, L. (1989). *Discipline review of teacher education in mathematics and science volume 1: Report and recommendations*, Canberra, Australian Government Publishing Service.

Suzuki, D. (1988). *Metamorphosis: Stages in a Life*, Sydney, Allen and Unwin.

Williams, B.R. (1979). *Education, training and employment: Report of the Committee of Inquiry into education and training*, Canberra, Australian Government Publishing Service, Australia.

Williams, R. (1990). 'Why science should be girl talk', *The Sunday Herald*, (Melbourne) 1 April, 15.

JOHN P. KEEVES[1] AND DIETER KOTTE[2]

7. PATTERNS OF SCIENCE ACHIEVEMENT: INTERNATIONAL COMPARISONS*

This chapter examines the patterns of gender differences in science achievement, attitudes and participation across several countries and over time. It argues that the patterns observed are the effects of societal forces and educational practices in schools that engender differences between boys and girls in attitudes, aspirations, expectations and levels of achievement. It demonstrates that the size of the observed differences between the sexes can be reduced and further that gradual changes are taking place, to varying degrees, in different countries. The issues are no longer whether greater equality between the sexes can be achieved, but where and how balance should be maintained in terms of principles of equity and social justice.

This article draws on the findings of the First and Second IEA Science Studies conducted in 1970–1971 and 1983–1984, respectively, by the International Association for the Evaluation of Educational Achievement (Comber & Keeves, 1973; IEA, 1988; Keeves, 1992; Postlethwaite & Wiley, 1991; Rosier & Keeves, 1991). Of the 19 countries that participated in the first study and the 23 that participated in the second study, 10 countries took part on both occasions. In addition, reference is made to related work carried out in Sweden (Duncan, 1989; Riis, 1991). Although other studies of differences between the sexes in science achievement, attitudes and participation have been carried out under other circumstances, the IEA reports all are concerned with large nationally representative samples, with common tests and analytical procedures. The IEA studies provide a rich source of evidence on which consideration of the issues addressed in this chapter can be based.

MORE EQUAL OPPORTUNITIES FOR MEN AND WOMEN

Although in many countries of the Western World women had gained the right to vote and access to education at all levels during the early part of the twentieth century, the different roles of men and women in society largely had remained unquestioned. However, since the late 1960s, many women have demanded the right to be treated as equals alongside men in most aspects of social, political and cultural life. As a consequence, the role of women in society has changed and educational opportunities for females have increased.

Entry into courses in preparation for many prestigious professional occupations requires advanced secondary education in mathematics and the physical sciences. Without competence in these subject areas, such educational opportunities, although accessible in principle, are denied in practice. Consequently, the lower levels of achievement by girls in mathematics and the physical sciences at the final stages of

77

L.H. Parker et al. (eds.) Gender, Science and Mathematics, 77–93.
© 1996 *Kluwer Academic Publishers. Printed in the Netherlands.*

secondary schooling are seen to restrict the educational and occupational choices available to them. The origins and the effects of such disparities in achievement warrant investigation so that a greater understanding is gained of the forces involved.

ACHIEVEMENT DIFFERENCES IN SCIENCE ACROSS GRADES OVER TIME

The conduct of the two related studies 14 years apart (i.e., 1970–1971 and 1983–1984) permitted an examination over time of sex differences in achievement not only between age groups, but also in the three major science fields of biology, chemistry and physics. Table I records the data for all 10 countries, and Figure 1 depicts the results averaged across countries.

The data are presented in terms of an *effect size*, which is a standardised measure of the difference in achievement between boys and girls, with allowance made for differences in the spread of scores between the different groups and subgroups. A positive effect size indicates that boys have higher average scores than girls. It should be noted that only rarely did girls outperform boys, with a minor exception occurring in biology. In general, these effects were larger in physics than in chemistry and larger again than in biology. Moreover, the effect sizes were larger at the terminal secondary school level, than at the 14-year-old level and larger again than at the 10-year-old level.

It is evident from Table I and Figure 1 that differences between boys and girls in average level of science achievement existed clearly for 10-year-olds, increased during the years of schooling, and differed across fields of science, being greater in the physical sciences than in biology. However, these results were biased by the effects of selection at levels of schooling beyond which the study of science or a particular field of science ceases to be mandatory. As a consequence, it is also necessary, particularly at the terminal secondary school level, to examine differences between the sexes in participation in the study of the different fields of science. In spite of the effects of selection, boys outperformed girls in science.

At the 10-year-old level there was little change over time in the sizes of the effects recorded. At the 14-year-old level, however, there was a significant reduction in the effect sizes in five countries, as well as when averaged across countries. The countries showing a reduction in effect size were Australia, Finland, Hungary, Japan and Sweden. England and Thailand were the exceptions. In Thailand, the sample employed at this level in the first study was known to be biased seriously; hence this comparison must be discounted.

At the terminal secondary school level, there was a significant reduction in the effects sizes in four countries, namely, Australia, England, Finland and Sweden. Three of these were common between the 14-year-old and the terminal secondary school levels. Furthermore, it should be noted that the significant falls in the effects sizes are associated with performance in the physical sciences, more particularly in physics.

These findings indicate that change in the relative differences in levels of achievement in science between boys and girls can take place over time. Such changes

TABLE I

Effect sizes* for sex differences in science achievement, 1970–71 and 1983–84.

Year Level	Subject	Study	Australia	England	Finland	Hungary	Italy	Japan	Netherlands	Sweden	Thailand	United States	Mean
10-Year-Old Level													
	Biology	FISS	•	-2	11	1	0	2	16	-4	-1	3	3
		SISS	8	5	-2	0	0	4	•	-10	•	14	2
	Chemistry s	FISS	•	20	31	12	9	12	31	14	9	13	17
		SISS	19	13	26	3	10	20	•	20	•	31	18
	Physics	FISS	•	47	48	37	38	24	56	50	40	34	42
		SISS	35	29	59	34	29	0	•	46	•	40	34
	Total	FISS	•	26	36	21	19	15	41	24	21	20	25
		SISS	25	21	34	12	15	12	•	26	•	35i	23
14-Year-Old Level													
	Biology	FISS	10	-13	15	3	11	43	11	18	-9	18	11
		SISS	16	25	16	4	14	28	17	13	22	32	19
	Chemistry s	FISS	23	0	20	11	12	34	11	11	-9	28	14
		SISS	21	22	8	1	21	33	32	18	30	33	22
	Physics	FISS	58	43	80	62	58	75	68	63	23	56	59
		SISS	30	40	53	31	49	27	66	41	41	39	42d
	Total	FISS	40	16	53	35	38	64	59	41	4	44	43t
		SISS	24d	34i	31d	17d	36	28d	49	25d	39i	40	32dt
Upper Secondary Level													
	Biology	FISS	20	17	38	15	10	•	44	47	3	36	26
		SISS	10	14	35	26	32	51	•	32	45	•	31
	Chemistry s	FISS	65	69	60	15	29	•	80	49	34	94	55
		SISS	35	39	45	37	28	62	•	51	26	•	40
	Physics	FISS	81	101	119	63	74	•	96	104	29	62	81
		SISS	68	75	87	57	72	82	•	85	53	61	71
	Total	FISS	72	82	94	41	50	•	95	86	29	56	64u
		SISS	46d	55d	58d	44	50	70	•	66d	46	43	51u

Notes:

FISS	First International Science Study		d	Decrease significant at the 5% level.
SISS	Second International Science Study		i	Increase significant at the 5% level.
•	No data collected.		u	Japan and Netherlands excluded.
s	Earth science included with chemistry.		t	Thailand excluded.

$$* \quad \text{Effect Size} = \frac{(\text{boys' score} - \text{girls' score})}{\text{pooled standard deviation across countries}}$$

Effect sizes x 100 recorded.

Effect sizes greater than .20 may be generally considered to be statistically significant at the individual country level, except where otherwise indicated.

Positive effect sizes indicate higher achievement by boys.

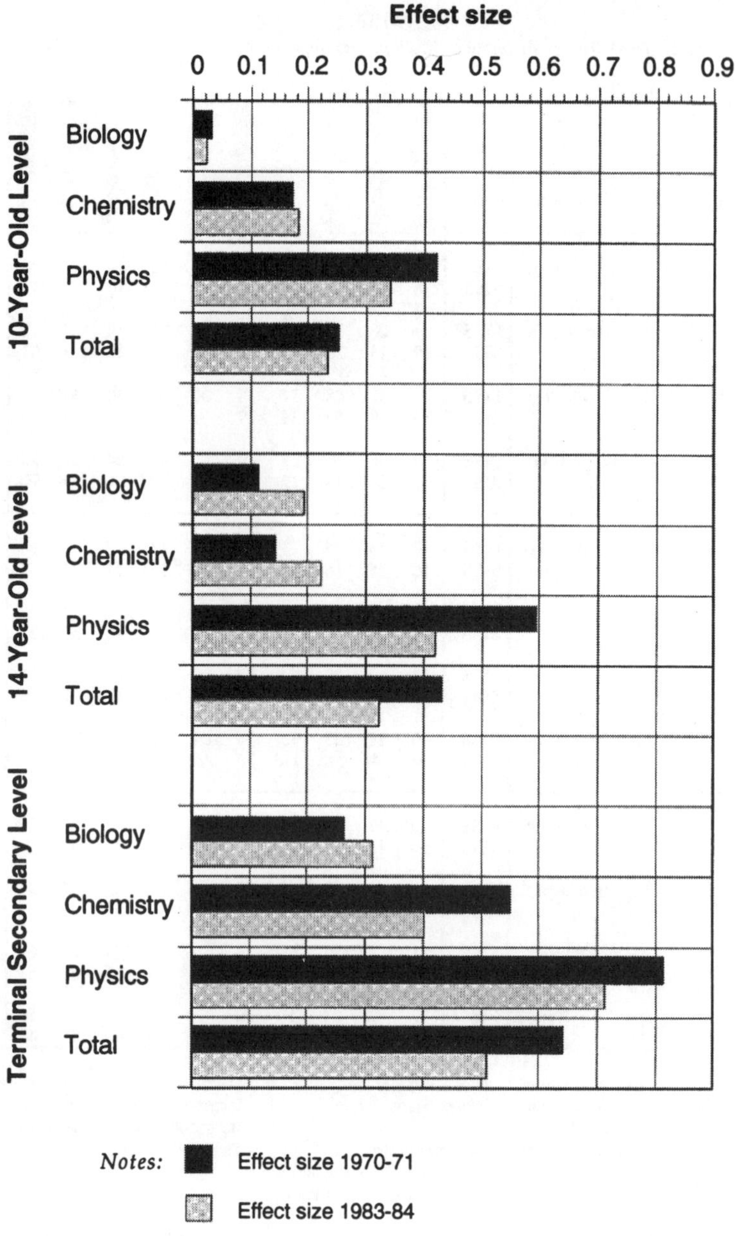

J.P. KEEVES AND D. KOTTE

Effect size

Figure 1.   Effect sizes for sex differences in science achievement averaged across 10 countries, 1970–71 and 1983–84.

suggest that the causes of such effects are not biological in origin but are associated with factors that involve both societal influences and educational practices. In Australia and Sweden, two of the countries where change occurred at both the 14-year-old and the upper secondary levels, special programs have been implemented in attempts to lift the performance of girls relative to boys in science. These programs would appear to have had the desired effects. Data clearly are needed to establish that these changes have been maintained and the gap further reduced over the decade from 1984.

ATTITUDES TO SCIENCE ACROSS GRADES

Table II gives information about five key attitudes: beneficial aspects of science; interest in learning science; ease of learning science; career interest in science; and school and school learning. It shows that for the four science-oriented attitudes, male students generally held more favourable attitudes. Even at the 10-year-old level, except in Italy, boys expressed greater interest in science and more favourable attitudes

TABLE II

Effect sizes* for sex differences in science attitudes, 1983–84.

Year Level / Attitude	Australia	England	Finland	Hungary	Italy	Japan	Netherlands	Sweden	Thailand	United States	No. of Items	Sign +	Sign –	Mean
10-Year-Old Level														
Importance of Science	-2	6	15	-11	-6	-4	•	20	•	-1	6	3	5	2
Interest in Science	2	8	8	11	-10	32	•	8	•	15	4	7	1	9
Ease of Learning Science	6	0	8	11	-4	9	•	s	•	16	2	5	1	8
School and School Learning	-36	-37	-45	-37	-30	-34	•	-36	•	-40	7	0	8	-37
14-Year-Old Level														
Beneficial Aspects of Science	11	30	-3	-10	6	25	22	22	14	5	7	8	2	12
Interest in Science	24	31	9	15	s	6	36	39	12	26	4	9	0	20
Ease of Learning Science	27	27	24	8	9	51	40	46	12	25	4	10	0	27
Career Interest in Science	25	21	6	6	17	40	24	18	14	10	7	10	0	18
School and School Learning	-23	-22	-41	-45	-33	-13	-14	-12	-31	-40	8	0	10	-27
Upper Secondary Level														
Beneficial Aspects of Science	14	16	18	10	8	45	•	43	10	20	7	9	0	20
Interest in Science	12	10	16	24	14	40	•	24	10	21	8	9	0	19
Ease of Learning Science	12	22	38	40	14	30	•	0	20	16	12	8	0	21
Career Interest in Science	34	32	36	18	43	51	•	49	12	31	7	9	0	34
School and School Learning	-22	-39	-32	-18	-26	-14	•	-8	-30	-30	7	0	9	-24

Notes: *Effect sizes x 100 recorded.

Positive effect sizes indicate more favourable attitudes of boys.

Effect sizes greater than 20 may be generally considered to be statistically significant at the individual country level.

s   Scale scores could not be calculated.

•   No data collected.

concerning the ease of learning science. At the 10-year-old level, there was little difference between the sexes in their views about the importance of science, except in Finland and Sweden where boys held more favourable views. At the 14-year-old level, only in Finland and Hungary did the girls hold more favourable attitudes than boys towards the beneficial aspects of science, and these differences were slight. For all other attitudes to science investigated at the 14-year-old level and in all other countries under survey, boys held more favourable attitudes than girls.

The second result, shown by Figure 2, is that differences between boys and girls in attitudes to science increased with age, particularly between the 10-year-old and 14-year-old levels. In general, this result parallels the increase in differences in science achievement between the sexes. However, it should be noted that the data recorded for attitudes to school and school learning did not follow the same pattern. First, it is the girls who, without exception, held more favourable attitudes to schooling. Second,

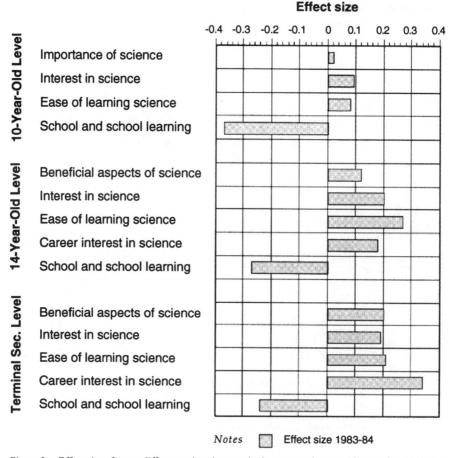

Figure 2. Effect sizes for sex differences in science attitudes averaged across 10 countries, 1983–84.

the average effect size decreased with age. These important differences between attitudes to science and attitudes to school and school learning – in terms of the direction of the gender effect, the change in magnitude with age, and the cross-national consistency of the results recorded – indicate that the effects under consideration are not artefacts which arise from the nature of attitude scale items and the use of Likert-type scales, as Riis (1991) has implied. Instead, they appear to represent strong and meaningful measures of underlying dispositions.

Further, these results would seem to indicate that attitudes to science and achievement in science may be linked together in the development of gender differences as students progress though the years of schooling.

PARTICIPATION IN THE STUDY OF SCIENCE

Throughout the years of lower and middle secondary schooling in most countries, most students study some science from all the different fields of science. However, as students move into the upper secondary school years, they are given freedom to choose which fields of science to study and whether or not to study science at all.

TABLE III

Percentage male students in science courses at the upper secondary school level, 1970–71 and 1983–84

Subject	Year	Australia	England	Finland	Hungary	Italy	Japan	Netherlands	Sweden	Thailand	United States
Biology	1970-71	45	45t	41	33	73	•	63	47	41	58
	1983-84	33s	41s	39t	41s	82	29s	•	57s	47	41s
Chemistry	1970-71	76	62t	62	41	75	•	71	60	41	55
	1983-84	57s	64s	47s	38s	70	73s	•	62s	47	57s
Physics	1970-71	78	70t	62	33	47	•	71	73	41	76
	1983-84	69s	77s	64s	45s	81	87s	•	71s	47	63s
Non-Science	1970-71	47	33t	23	80	64	•	57	38	33	41
	1983-84	45	45	•	23	45	33	•	33s	41s	45

Notes: s   Figures for 1983-84 which are derived from numbers of male and female students in science classes obtained from the students and the schools are more accurate than the numbers of male and female students from the proportions of male students in the sample.

          t   Official figure.

          •   No data collected.

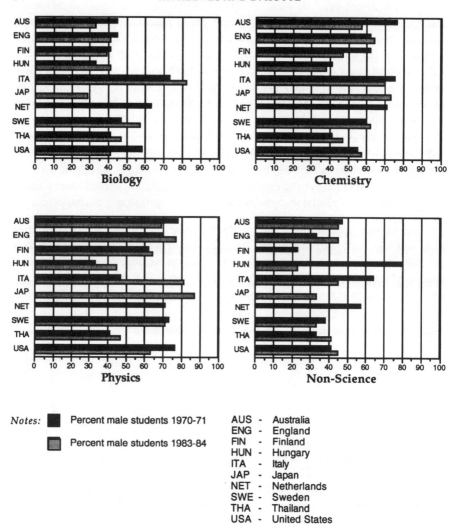

Figure 3. Percentage male students in science courses at the upper secondary school level, 1970–71 and 1983–84.

Under these circumstances, the tendencies on the part of girls to perform less well than their male counterparts in the physical sciences, to view the learning of science to be more difficult and to express less interest in the study of science are likely to influence their decisions to cease taking science, or to study a particular field of science. The proportions of those at school studying science or a particular field of science also are influenced by differences between the sexes in retention rates, and by requirements imposed by matriculation examinations or the conditions for entry into higher education. Although this situation is a complex one, both within and across countries, relatively consistent patterns of results can be observed.

Table III records the estimated proportions of male students studying in each field of science at the upper secondary school level for both 1970–1971 and 1983–1984. These data are presented graphically in Figure 3. With the exception of Italy on both occasions, the study of biology is preferred by girls rather than boys. However, the study of chemistry is largely a male preference, although Hungary and Thailand are the exceptions on both occasions. With the exception again of Hungary and Thailand, the study of physics is dominated strongly by males. These results suggest that differences between boys and girls in their relative levels of performance in the different fields of science influence, in part, their decisions about whether or not to continue with the study of the subject after certain matriculation requirements have been taken into account.

With few exceptions, the evidence presented in Figure 3 indicates that there has been little change in the differences between boys and girls in their tendencies to choose to study one field of science in preference to another between 1970–1971 and 1983–1984. Thus, although there are many factors operating to influence participation rates in the different fields of science, there was a strong tendency for male students to take the physical science subjects and for female students to take the biological science subjects, but there was little change in the relative proportions over time.

CHANGES IN YIELD

The complex relation between retention rates and participation in the study of the particular fields of science confounds the data on changes in differences between the sexes in the proportions studying each field and in levels of achievement. The issues to be addressed must be considered in terms of 'how many get how far?' with respect to both performance in science and the differences between male and female students. A simple index which takes into account both the proportion of the age group and level of achievement is that of *yield*. A simple estimate of *yield* is given by the product of achievement and the proportion of the age group involved. Table IV records, for both 1970–1971 and 1983–1984 and for male and female students separately, the proportion of the age group at school, the levels of achievement of the samples, and the estimates of the yield index.

In the penultimate section of Table IV, there is recorded, for both occasions, information about the retention rates, which are the proportions of the age group enrolled in full-time schooling on both occasions and the proportions under survey who were enrolled in courses that would permit them to continue with further study at the post-secondary level. In general, a marked increase in the proportion of the age group enrolled at school occurred between occasions.

However, no increase took place over time in England, and a slight drop occurred as a result of the restructuring of upper secondary schooling in Sweden. In Hun-

TABLE IV

Yield coefficients for male and female students at the upper secondary school level, 1970–71 and
1983–84

Year Group	Australia	England	Finland	Hungary	Italy	Japan	Netherlands	Sweden	Thailand	United States
1970-71										
Males										
Proportion of an Age Group	32	20	16	24	18	•	16	31	7.5	74
Mean	28.0	27.8	25.2	25.6	17.6	•	26.7	25.6	12.9	16.5
Yield	890	560	400	610	320	•	430	790	100	1220
Females										
Proportion of an Age Group	26	20	26	32	14	•	10	29	12.5	76
Mean	20.8	19.6	16.1	20.9	13.5	•	17.5	17.0	12.1	11.0
Yield	540	250	420	670	190	•	180	490	150	840
1983-84										
Males										
Proportion of an Age Group	36	20	31	13	32	60	•	28	12	75
Mean	18.0	21.0	19.6	23.3	13.3	20.0	•	19.6	11.2	12.2
Yield	650	410	610	300	430	1210	•	550	130	910
Females										
Proportion of an Age Group	42	20	49	23	36	66	•	20	16	85
Mean	15.1	17.4	15.8	20.5	10.1	14.6	•	15.3	8.1	9.2
Yield	640	350	780	480	360	960	•	430	130	780
Ratio Male/Female Yields										
1970-71	1.6	2.2	1.0	0.9	1.7	•	2.4	1.6	0.7	1.5
1983-84	1.0	1.2	0.8	0.6	1.2	1.3	•	1.3	1.0	1.2
Retention Rates (% of Cohort)										
1970-71	29	20	21	28	16	70	13	30	10	75
% in School 1983-84	39	20	59	40	34	89	14	28	29	80
% under Survey 1983-84	39	20	41	18	34	63	•	28	14	80
% Male Students										
1970-71	58	58	44	44	57	52	64	52	52	52
1983-84	47	50	39	33	47	50	•	50	44	47

Note: •   No data collected.

gary, while the proportion enrolled at school increased between 1970–1971 and
1983–1984, less than half of those enrolled on the latter occasion were in academic
schools, and able to continue with university courses, the remainder being enrolled
in vocational schools. The final rows of Table IV record the percentage of male
students at the upper secondary school level. Greatly increased retention of girls
relative to boys is evident between occasions.

In examining Table IV, it should be noted that the science achievement test
scores were based on different tests on the two occasions and direct comparisons of

either achievement or the yield indexes cannot be made readily. Furthermore, the scores recorded are concerned with the performance of students whether or not they were studying science. As a consequence, the index of yield applies to all students completing secondary schooling. Some of these students had not specialised in the study of science, or a field of science, and many students had not been required to study science during their final year of schooling. This seemed the only fair way to estimate meaningfully the yield of the school system with respect to science learning, and to make allowance for the different proportions of male and female students involved.

From Table IV, the cases of Finland and Hungary, where the ratio of the male/female yield index has fallen well below unity during 1983–1984, warrant closer examination. It appears that, in both of these countries, boys opted out from formal academic education, preferring to move more rapidly towards employment and the earning of money, through a shorter educational program in vocational schools. As specific vocational skills become obsolete and more difficult to define, the more general education provided in academic schools appears to have increasing value. Consequently, the question must be raised as to whether it is becoming necessary to provide boys within academic schools with more effective programs that are more relevant to these male students' long-term needs. In all other countries where secondary schooling is now completed by greater numbers of girls than boys, it would seem necessary to monitor carefully the trends recorded between 1970–1971 and 1983–1984 and beyond.

ENGENDERING SCIENCE LEARNING

During the years of schooling, the differences between boys and girls in their achievement and attitudes to science increase more in the physical sciences than in the biological sciences. Thus, it is not surprising that more boys than girls continue with the learning of physics and chemistry when the study of science ceases to be mandatory. Commonly, although not universally across countries, a greater number of girls study biology rather than physical sciences, even though generally girls did less well than boys on the biology tests which were administered in the IEA studies. These findings are of no surprise to English-speaking science educators. What is of interest is the common pattern of results across the 10 countries.

The evidence presented in this chapter shows that, in some countries, significant changes occurred in the science achievement of girls relative to boys over the period from 1970 to 1984. Moreover, as a result of the increased retention rates for girls at the upper levels of secondary schooling, the ratios of the yield indexes for boys compared to girls have been reduced towards a value of unity and in two countries have fallen well below unity. These changes suggest that societal and educational policies and practices which have been introduced since the 1970s, at least in part, could be having their desired effects. Nevertheless, the processes by which science achievement, attitudes and participation patterns are formed, or can be changed, are not understood clearly. Since the early 1970s, when the analyses of data collected in the first IEA science study were carried out, new analytical techniques have been developed for investigating such processes. One specific goal in these analyses has been to examine the manner in which certain factors mediate between the sex of student and the outcome measures of achievement and future participation in the study of science. Alternatively, the analyses could indicate that the processes of mediation are very different, and therefore that separate analyses must be carried out in order to clarify the way in which certain factors influence achievement and participation. Such mediation processes could operate at the individual student, the school and/or the classroom group levels.

The analysis of data can be carried out at three levels – between students within groups, between groups, and between students across groups. Each approach to analysis has its shortcomings. Only by undertaking multilevel analyses in which effects were examined at the between student within group and the between group levels simultaneously could the effects of bias be eliminated. However, where there is no variation within a group on such a variable as sex of student, which occurs in single-sex schools, these schools are necessarily dropped from the analysis. This demands that separate analyses are carried out for boys' schools, girls' schools and coeducational schools, to examine whether different processes are in operation in different types of schools (Kotte, 1992). Alternatively, separate analyses can be carried out for boys and girls, and the effects associated with either single sex or coeducational schooling can be estimated.

THE MEDIATING EFFECTS OF ATTITUDES AND VALUES ON SCIENCE ACHIEVEMENT

Table V records the relationships associated with the variable sex of student and science achievement at the three different age levels. The analyses were conducted at the level of between students across groups. Table V gives several measures of the relationships involved. First, the zero-order correlation is recorded, to provide a measure of the association between sex of student and science achievement. Second, the direct effect is reported to represent the relationship between sex of student and science achievement after the influence of other factors has been taken into account. Third, the indirect effect of sex of students on science achievement is provided to indicate the extent to which the influence of sex of student is mediated through other factors such as attitudes and values. Finally, the total effect of sex of student is given. This combines together the direct effects and the mediated effects, to provide a measure of the effect of sex of student acting directly and indirectly to influence science achievement, after the other variables included in the model have been taken into consideration.

The results are consistent across the three age levels and estimates of effect in excess of 0.05 (note decimal points are omitted in Table V) are considered significant. In all countries, at all three age levels, the direct effects of sex of student on achievement were significant after other factors had been considered. Only in seven out of

TABLE V

Effects of sex of student on science achievement and future participation in science for between students within classroom, 1983–84

Year Level / Comparison*	Australia	England	Finland	Hungary	Italy	Japan	Netherlands	Sweden	Thailand	United States	Sign +	Sign –	Mean
Criterion - Science Achievement													
10-Year-Old Level / Between Students Within Country													
Correlation	-13	-10	-18	-8	-8	-4	•	-13	•	-17	0	8	-11
Direct Effect	-14	-11	-18	-11	-10	-6	•	-11	•	-16	0	8	-12
Indirect Effect	0	0	+1	-1	+1	-2	•	-1	•	-1	2	5	0
Total Effect	-14	-11	-17	-12	-9	-8	•	-12	•	-17	0	8	-12
14-Year-Old Level / Between Students Within Country													
Correlation	-13	-17	-17	-7	-18	-15	-24	-14	-21	-21	0	10	-17
Direct Effect	-16	-15	-18	-13	-18	-11	-25	-11	-23	-20	0	10	-17
Indirect Effect	0	+1	0	+5	+1	-5	0	-8	+1	0	4	2	0
Total Effect	-16	-14	-18	-18	-17	-16	-25	-19	-22	-20	0	10	-17
Upper Secondary Level / Between Students Within Country													
Correlation	-23	-30	-37	-24	-26	-36	•	-32	-20	-27	0	9	-28
Direct Effect	-12	-13	-27	-12	-25	-9	•	-18	-17	-19	0	9	-17
Indirect Effect	-3	-3	-9	-6	-1	-1	•	-16	-1	-5	0	9	-5
Total Effect	-16	-16	-36	-18	-26	-10	•	-32	-18	-24	0	9	-22
Criterion - Future Participation in Science													
14-Year-Old Level / Between Students Within Country													
Correlation	-7	-3	0	+8	-12	•	-8	-12	+2	0	2	6	-4
Direct Effect	0	0	0	0	-11	•	0	0	0	0	0	1	-1
Indirect Effect	0	0	0	0	+1	•	-10	0	0	0	1	1	-1
Total Effect	0	0	0	0	-10	•	-10	0	0	0	0	2	-2
Upper Secondary Level / Between Students Within Country													
Correlation	-4	-11	-12	-31	-7	-31	•	0	0	-10	0	7	-12
Direct Effect	0	0	0	-14	0	0	•	8	0	0	1	1	-1
Indirect Effect	0	-5	-12	-11	0	-11	•	-8	0	-8	0	6	-6
Total Effect	0	-5	-12	-25	0	-11	•	0	0	-8	0	5	-7

Notes: *   Decimal points omitted
        •   No data collected.

27 analyses were the indirect or mediating effects of sex of student in excess of 0.05 considered significant. The largest effects of sex of student mediated through attitudes and values occur at the 14-year-old and upper secondary school levels in Sweden and at the upper secondary school level in Finland.

In all but one case, the significant indirect effects of sex of student enhance the achievement of boys. In the one contrary case in Hungary, in which indirect effects raise the performance of girls, these effects are marginal. Larger indirect effects of sex of student being mediated through attitudes and values were recorded at the upper secondary school level than at the 14-year-old and 10-year-old levels. When added to the direct effects, these indirect effects raise the value of the estimates of total effects, but still there is little evidence (except in Sweden and Finland at the upper secondary level) of substantial influence of sex of student being mediated through attitudes and values. Thus, there is little evidence to indicate that the processes which operate to generate the increased differences across age levels in science achievement are working through the attitudes and values towards science that were investigated in the Second IEA Science Study. It is possible, however, that other attitudes and values might operate to mediate between sex of student and achievement in science.

Table V also records the measures of effect of sex of student on future participation in science courses at both the 14-year-old and the upper secondary school level. At the 14-year-old level, factors other than sex of student influenced whether or not a student continued with the study of science. Only in Italy and The Netherlands were significant total effects recorded, and only in The Netherlands was there evidence of recognisable indirect effects operating through attitudes and values. In Italy the effects were largely direct.

At the upper secondary school level, there was evidence in six of the nine countries of sex of student operating indirectly through other variables to influence the proposed future study of science. Only in two countries were these direct effects of significance. In Hungary, the direct effect enhanced the involvement of boys, and in Sweden the direct effect enhanced girls' future participation in science courses.

The analyses which were carried out to examine the processes by which sex of student influenced science achievement and future participation in science courses recorded effects that were weaker than had been expected. It had been hypothesised that attitudes and values towards science would play a significant part in mediating the effects of the variable of sex of student. In general, such a mediating role for attitudes and values towards science was small, although it was an influence on future participation at the upper secondary school level in certain countries. It would appear that other factors were operating to mediate the effects of sex of student on the outcomes of science achievement and future participation.

In a study of factors influencing science achievement among boys and girls at the middle secondary school level in the developing country of Botswana, Duncan (1989) found it necessary to undertake separate analyses for girls and boys when testing the complex model that she hypothesised. She showed that gender typing of science and gender typing of science in the classroom had small but recognisable influences on science achievement and participation for both boys and girls. The direction of the relationships, however, differed between the sexes. For boys, the gender typing of school science had a positive effect; for girls, it had a negative effect. Nevertheless, among both boys and girls, gender typing acted indirectly upon achievement through its effects on achievement-related attitudes. Among boys, gender typing fostered the idea that science was an easy subject, and that it was useful for them. These ideas, in turn, increased boys' liking for science, and subsequently their achievement and further participation. The opposite process operated for girls.

In addition, the tendency to perceive science as male was influenced indirectly by gender-role ideology in the case of boys and directly in the case of girls. Gender-role ideology also influenced occupational gender typing. Thus, girls and boys who supported more strongly the self-image linked to their own sex were more likely to endorse the view that men and women should engage in different types of occupations. These results provide at least partial support for the generally accepted view that one of the reasons for the low level of participation of girls in the physical sciences involves gender stereotyping. Thus, girls who see science in general and science in the classroom as a male domain of activity are likely to have less positive attitudes to science and to perform less well.

The influence on science performance of gender typing of science was more evident in Duncan's study among boys than girls, and boys revealed a greater tendency to see science and school science as male. Boys who saw school science as masculine had more positive attitudes to science and achieved better in science than those who did not. Thus there was clearly a network of effects through which gender typing operated to influence performance, but, although the magnitudes of these effects were statistically significant, they were not large. Moreover, while these processes were tested at the secondary school level, where the evidence available indicated that such processes were most likely to be in operation, the analyses did not establish that they were influenced by forces at work within the schools and classrooms. The school, the classroom and the peer group would appear to be the most likely places for such developments to occur. However, to test such hypothesised relationships, a study would need to be designed specifically and the data subsequently analysed in ways that would permit relationships at both the student and classroom levels to be tested simultaneously in a path model. Appropriate analytical procedures have only recently become available. Nevertheless, this study undertaken by Duncan draws attention to the need for methods of analysis with the capacity to examine the manner in which mediation processes operate through gender typing and gender-role ideology to influence educational outcomes.

CONCLUSION

On the basis of evidence gathered mainly from the First and Second IEA Science Studies, this chapter has demonstrated that, in both developed and developing countries, marked patterns of gender differences in science emerge by the end of secondary schooling. It has explored some of the details of these patterns of participation, achievement and attitudes to science, with particular reference to data from the 10 countries which took part in both studies. It has shown that girls made some significant gains in science achievement between 1970–1971 and 1983–1984 in some countries. However, overall, on the tests, boys' average level of science achievement was higher and their overall attitudes to science were more positive than those of girls at age 10 and the differences between the sexes increased as the students progressed through secondary school. These differences were especially evident in the physical sciences, which also were the science subjects where boys predominated in the upper levels of secondary schooling.

While greater understanding is needed of how differences arise between boys and girls in science achievement, attitudes and participation, there is no doubt from the evidence about differences in retention rates and science achievement presented in this chapter that changes can be brought about if appropriate efforts are made. However, the findings documented in this chapter for Finland and Hungary show a trend over time in which significant numbers of boys are withdrawing from academic education to take vocationally-oriented programs. This would appear to be occurring at a time when most developed countries are experiencing serious shortages of persons with higher-level skills in the fields of science and technology.

Two very different approaches have emerged from the evidence presented in this chapter that might be pursued to overcome these shortages. First, the *yield* from the school systems of most countries, when both achievement in science and retention rates are taken into consideration, has increased substantially from 1970 to 1984 for girls relative to boys. As a consequence, there is now a sizeable reserve of scientific competence among girls that is not being fully utilised as girls turn away from courses that would prepare them for scientifically-based professional occupations following the completion of secondary schooling. It is clear that greater efforts should be made to attract girls into science-type courses, particularly those related to mathematics and the physical sciences during the final years of secondary schooling and in the transition from school to higher education.

Second, the data presented suggest that boys are less inclined to remain at school to obtain the benefits of 12 years of education. They have turned instead to employment, vocational training and short-term financial rewards. Thus, there may be a need to provide, within upper secondary education, courses that are more attractive to boys, particularly courses related to science and technology that would help to overcome the present shortages of highly skilled personnel. Finland and Hungary show trends that are in danger of becoming common in other parts of the world. These two countries seem to have the greatest need for reform in upper secondary education, but other countries appear to have similar problems.

It is important that the developments investigated in the First and Second IEA Science Studies be monitored in the future. In addition, it is important that implementation of policies which are concerned with the increased opportunities for women in education and society at large should be maintained vigorously.

[1]The Flinders University of South Australia, Australia;
[2]University of Hamburg, Germany

NOTES

*The term 'sex' in this article refers to the observable biological differences between male and female students. The term 'gender' is used to denote 'the set of meanings, expectations and roles that a particular society associates to sex' (Megarry, 1984). Hence, where differences between males and females in achievement and attitudes are considered, the term 'sex' is used; where reference is made to science programs and science teaching which are socially determined, the adjective 'gender' is used.

REFERENCES

Comber, L.C. & Keeves, J.P. (1973). *Science education in nineteen countries*, Stockholm, Almqvist and Wiksell.

Duncan, W. (1989). 'Engendering school learning: Science attitudes and achievement among girls and boys in Botswana', *Studies in Comparative and International Education*, No. 16, Stockholm, Institute of International Education, University of Stockholm.

IEA (International Association for the Evaluation of Educational Achievement) (1988). *Science achievement in seventeen countries: A preliminary report*, Oxford, Pergamon Press.

Keeves, J.P. (ed.) (1992). *The IEA study in science III: Changes in science education and achievement, 1970–1984*, Oxford, Pergamon Press.

Kotte, D. (1992). *Gender differences in science achievement in 10 countries: 1970/1971 to 1983/1984*, Frankfurt/Main, Germany, Lang.

Megarry, J. (1984). 'Sex, gender and education', in S. Acker (ed.), *Women and education*, World Yearbook of Education, London, Kogan Page, 14–28.

Postlethwaite, T.N. & Wiley, D.E. (1991). *The IEA study in science II: Science achievement in twenty-three countries*, Oxford, Pergamon Press.

Riis, U. (1991). 'Girls in science and technology', in T. Husén and J.P. Keeves (eds.), *Issues in science education: Science competence in a social and ecological context*, Oxford, Pergamon Press, 109–122.

Rosier, M.J. & Keeves, J.P. (1991). *The IEA study in science I: Science education and curricula in twenty-three countries*, Oxford, Pergamon Press.

GILAH C. LEDER

8. EQUITY IN THE MATHEMATICS CLASSROOM:
BEYOND THE RHETORIC

(T)he Australian colonies were among the earliest to emancipate women politically, to offer girls equal educational opportunities, and along with other pioneering societies of the American and Canadian West, and New Zealand, they were seen as promoting female behaviour and personality which allowed assertiveness, friendliness and independence. (Grimshaw, 1982, p. 3)

For more than a century, there have been no formal barriers preventing Australian women from participating in all levels of education – whether primary, secondary, or tertiary. Indeed, in recent years, more females than males have remained in full-time education until the end of secondary schooling in Australia (as well as in countries such as the United Kingdom).

Yet a number of consistent, though relatively small, gender differences in participation and performance in mathematics still are described in the literature (Leder, 1992). Briefly, differences in participation continue to be observed in higher-level, more intensive mathematics courses and related applied fields. For example, 1993 data from an examination taken at the end of secondary school (the Victorian Certificate of Education (VCE) examination) indicated that more females than males registered for the mathematics unit Change and Approximation, more males than females enrolled in the mathematically more demanding Extensions unit, while approximately equal numbers of females and males sat for the mathematics unit Space and Number. Generally there is much overlap in the mathematical performance of females and males. When significant differences in performance occur, they tend to favour males, particularly on traditionally timed tasks found in tests and examinations, but not necessarily on longer-term project assignments or in-class work. At the same time, there are often subtle differences in the ways in which females and males regard themselves, and are regarded by others, as learners of mathematics. Students capable of continuing with mathematics, but who believe the subject to be inappropriate for them, are more likely to self select out of mathematics courses and hence out of other areas for which such work is a prerequisite.

Many explanations have been put forward to account for the observed gender differences in mathematics learning. Those working within feminist frameworks argue strongly that gender is an important influence on the way in which society is structured. Traditionally, males' beliefs, experiences and behaviours have been accepted as the norm. Elsewhere (Leder, 1986, 1992), I have argued about the desirability of concentrating on variables of particular relevance to educators. These include environmental variables (situational, personal and curriculum) as well as ones associated with the learner (cognitive and psychosocial). Of particular interest for the purposes

95

L.H. Parker et al. (eds.) Gender, Science and Mathematics, 95–104.
© 1996 *Kluwer Academic Publishers. Printed in the Netherlands.*

of this chapter are explanations that focus on the ways in which females and males are treated and behave in mathematics classes.

INSIDE THE CLASSROOM

Teachers play a crucial role in interpreting and implementing policies adopted by or imposed on schools. Many government initiatives assume that schools can improve life conditions and choices for girls and – in the long term – for women. However, as Sampson (1989) has pointed out:

This is not one of the original purposes of schools, and they are finding it difficult. Trained to impart literacy, numeracy and essential training and socialisation for work and citizenship, teachers have little or no experience in helping girls to overcome such handicaps as sex-stereotyped subject choices or ambivalence about possible future roles as a single or family breadwinner. (p. 2)

Whatever the formal organisation of the setting in which mathematics is taught, the general climate inside the mathematics classroom is likely to reflect that of the school and the broader society in which it is set.

Teachers' in-class behaviours have been described using a great variety of observation schedules. Many of these rely on low inference observation schedules of classroom events rather than on more qualitative and interpretative data-gathering techniques. Typical of the conclusions drawn are those of Harris and Rosenthal (1985) who reviewed more than 130 studies concerned with expectancy effects:

Teachers who hold positive expectations for a given student will tend to display a warmer socio-emotional climate, express a more positive use of feedback, provide more input in terms of the amount and difficulty of the material taught, and increase the amount of student output by providing more response opportunities and interacting more frequently with the student. (p. 377)

Particular attention has been focussed on the ways in which teachers interact with the female and male students in their classes, and the effects of these behaviours on the learning of their students. Common findings have been summarised well by Eccles and Blumenfeld (1985).

We like many others, have found small but fairly consistent evidence that boys and girls have different experiences in their classrooms. . . . (W)hen differences occur, they appear to be reinforcing sex stereotyped expectations and behaviours. (p. 12)

It is worthwhile to look in some detail at the findings of an Australian study in which 32 grade 3 to 10 classes were surveyed during mathematics lessons (Leder, 1987, 1988, 1993). The total sample consisted of almost 600 students and contained approximately equal numbers of boys and girls.

AUSTRALIAN DATA

The methodology employed built on that used in previous work. Two quite distinct observation schedules were used to capture as fully as possible the rich classroom environment in which teachers and students interact. No restrictions were placed on teachers' presentations or behaviours, apart from a request to include as much oral and questioning work as possible in the lessons to be monitored.

The lessons of interest were videotaped and subsequently analysed. This approach allowed a complex mix of behaviours to be described and quantified without needing to rely on a conspicuously large number of observers. Furthermore, all interactions between teachers and every student in the class could be noted. The procedure used for data collection thus was particularly efficient and, according to the teachers observed, surprisingly unobtrusive.

The two observation schedules used to quantify teachers' behaviours relied on the methods advocated by Brophy and Good (1970) and Rowe (1974a, 1974b). The former focusses on teacher-child dyadic interactions, and the latter on wait-time and length of interaction measures. Both schemes have been used widely in other classroom observation studies.

Specifically, interactions were categorised according to type of question (high or low level cognitive question), nature of other oral exchanges (procedural, discipline or subject matter/work related), type of exchange (single or multiple/sustained), setting of the exchange (public or private) and initiator of the exchange (teacher or student). For questions, the wait-time (i.e., the time interval that began when the teacher stopped talking after posing a question and terminated when the student called on responded or the teacher spoke again) was measured. The engagement time between the teacher and the student called on, as well as the (mean) engagement time per single exchange, also were noted.

Details of the various categories are described in Leder (1993). Briefly, low cognitive questions referred to routine, rather simple recall, questions such as $8 + 9 = ?$ High cognitive level questions placed a higher cognitive demand on the student. Some synthesis, generalisation or abstraction usually was required. For example, a student could be asked to compare two answers and justify the selection of the alternative chosen. Exchanges about routine matters ('use a ruler for that', 'start on a new page') were coded as procedural. Discipline exchanges are self explanatory. Teacher explanations and attendant student questions about the mathematics subject matter at hand were coded as work-related exchanges.

The stop watch built into the camera facilitated gathering of the time-related data. Repeated analysis of selected segments revealed a high inter- and intra-coder reliability for each of the categories of interest.

INTERACTIONS BETWEEN TEACHERS AND STUDENTS

It is convenient to report sequentially the results obtained with the two different observation schedules. Conclusions, however, should be drawn from the total picture described on the basis of the more than 10,000 interactions monitored.

The frequency-of-interaction data were consistent and striking. At each of the four grade levels observed (grades 3, 6, 7 and 10), on average boys interacted more frequently overall with their teachers than girls. In grades 3 and 6, boys had approximately 20% more interactions with their teachers in mathematics classes than did girls. At the grade 7 level, the discrepancy in favour of boys was somewhat less: for every nine interactions with girls, there were approximately 10 with boys. The greatest differences occurred at the grade 10 level where, on average, boys had 25% more interactions with their teachers than did girls. Despite variations in the behaviours of individual teachers, the global data reflect the climate in the majority of the classes observed.

Substantial differences also were found in the interactions between teachers and girls as a group and boys as a group in the different categories observed. For example, boys tended to be asked somewhat more questions than girls, and to have more discipline exchanges, more work-related exchanges, more public interactions, and more teacher-initiated interactions. While not all the observed differences were found to be statistically significant, there was a remarkable consistency in favour of boys for both the statistically significant and nonsignificant findings. It is worth stressing that, when all discipline exchanges were discounted, boys still were found to have more interactions than girls with their teachers.

Interestingly, the consistent bias that characterised the frequency of interaction data was not found when the length-of-time observation schedule was used. With the latter scheme, statistically significant findings (12) were outnumbered by non-significant differences (24). Eight of the significant differences related to the low and high cognitive question categories. Above grade 3, where differences occurred, boys appeared to be given more time on high cognitive level questions, whereas girls had more time on routine low level questions. The former, it could be argued, form a better preparation than the latter for advanced and intensive mathematics courses. The observation that, at each grade level, students were asked far more low than high cognitive level questions offers a useful measure of the quality of the learning environment in the average mathematics classroom.

It is inappropriate to draw conclusions from the length of interaction time data if considered in isolation. However, the slight and subtle differences noted in the patterns of interactions between teachers and boys as a group and girls as a group seem to reinforce the more striking differences captured by the frequency of interaction observation schedule. Taken together, the two sets of data suggest that boys and girls are treated differently by their teachers in mathematics classes. They also highlight the extent to which descriptions of gender bias within the classroom are influenced by the data gathering technique employed.

The differences in teacher treatment of girls and boys preceded gender differences in mathematics performance. As part of the study, two measures of mathematics achievement – the appropriate level of the Operations Test in the Mathematics Profile Series (Australian Council for Educational Research, 1977) and a teacher ranking – were obtained for each student. No differences were found in the mathematics achievement of girls and boys except at the grade 10 level, where boys performed better than girls. Yet data from this and other studies (Leder, 1989; Leder et al., 1994) confirmed a number of differences in the ways in which boys and girls with comparable achievement in mathematics perceived themselves, on average, as learners of mathematics.

To investigate students' mathematics-related attitudes, almost 100 students (43 girls and 51 boys) in grades 3 and 6 were interviewed, in a one-to-one setting, about work covered in recently-completed mathematics lessons. Issues relating to mathematics more generally also were discussed.

The interview data revealed a number of differences in the ways in which boys and girls responded to questions not specifically related to the mathematics work discussed in class. For example, approximately 25% of the boys compared with 14% of the girls considered themselves to be above average in mathematics. In contrast, 10% of the boys, but approximately one quarter of the girls, thought themselves to be below average in mathematics. Yet, according to teacher assessment and commercially-designed mathematics tests, there were no differences in the performance of boys and girls in any of the classes tested.

Half the boys (51%) nominated mathematics as their favourite subject, and this was followed by sport (18%) and language (14%). Girls, however, were more likely to nominate language (42%), with mathematics and sport preferred by 21% and 9% of girls, respectively. While it is important not to ignore the considerable overlap in the responses of the boys and girls, the differences in the answers given by the two groups are noteworthy and apparently persistent. In a recent survey of a much larger sample of 1,700 students from 20 schools in grades 7 to 10 (Leder *et al.*, 1994), sport was nominated as the favourite subject by 37% of the boys and 19% of the girls, mathematics by 10% and 7%, respectively, English by 6% and 13%, respectively, while only a modest 5% and 4%, respectively, indicated science to be their favourite subject. Collectively, the responses suggest that the broad context in which mathematics is learnt differs subtly for boys and girls.

Other differences also were found. For example, one third of the girls, compared with just over 10% of the boys, claimed that they liked doing easy mathematics; 26% of the girls and 45% of the boys indicated a preference for doing 'difficult things' in mathematics. Previously, there was noted a slight tendency for boys to be given more time by teachers on high cognitive level questions, and girls to be given more time on the less-taxing low-level questions. Are students perhaps conforming to teacher expectations? Or are teachers aware of the student preferences and reacting to them? Further examination of mathematics lessons, using a more open, qualitative approach, could provide some useful insights.

Mathematics lessons in American schools, it has been claimed, frequently conform to a recognisable format:

While the profile and interpretation of mathematics teaching . . . is neither simple nor consistent, the predominant pattern is extensive teacher-directed explanations and questioning followed by student seatwork on paper-and-pencil assignments. (Romberg, 1988, p. 3)

The generalisability of these conclusions is confirmed by observations of Australian classrooms. Two extracts are used to illustrate the most common learning climate we observed.

Lesson 1

The first lesson was concerned with fractions. It began with activity involving the whole class. Groups of numbers were written on the board and students were asked to identify those which were 'different'. These, they were reminded, were called fractions. The features of a written fraction were explained.

A mandarin-orange was used to demonstrate the concepts of 'whole' and 'equal parts'. Students were asked to hold up specified numbers of the 10 segments into which the mandarin-orange had been divided. The fractions thus illustrated were named and written on the board. This procedure was used to demonstrate, for example, that three of 10 equal parts could be written as 3/10.

Other shapes (circles, squares) subsequently were drawn on the board and divided into varying numbers of equal parts. Students again were called on to shade certain parts, for example, six of eight equal parts. Each time, the corresponding fraction was written on the board: 'He is shading six [6/] out of the eight [6/8] equal parts'.

The next part of the lesson required the students to fold strips of paper according to the teacher's verbal directions. She demonstrated the folding techniques to be used, and then asked students to mark certain sections and to write the appropriate fraction on the paper. The fraction nominated also was written on the board.

Finally, attention was drawn to a series of activities to be done by the students themselves. Their requirements were discussed carefully and the class then was set to work on these tasks. The high level of teacher direction in this lesson is clearly apparent.

Lesson 2

The second lesson, also typical of many which we observed, was concerned with division. The aim was to describe the formal procedure for dividing with carrying (e.g., 620 divided by 5).

The lesson began with revision of simple division sums already familiar to the students. This was followed by a demonstration of the desired procedures through working a number of examples – using some student involvement – on the board. The students then were asked to complete the remaining examples which had been written on the board. If they wished, they could use concrete materials, which were readily available, to help them in their workings of the sums. Some supervision by the teacher of student work occurred during this time. Towards the end of the lesson, the students were invited to work 'on the floor with Mrs A' if they were experiencing difficulties with the work at hand. Several students took up the offer.

To convey as faithfully as possible the tone and atmosphere of the lesson, the teacher's explanations used in the early part of the lesson are reproduced verbatim below:

During the year, we have looked at plus sums, addition sums, subtraction sums, and we've looked at multiplication or times sums. We've looked at them using small numbers and big numbers and we've done them as grown up people or people from grade 3 up have done. We're going to go over again, today, past what we've done . . . I've said to you . . . nine, how many three's? In your mind you've been able to tell me. . . .

Tell me the answer: nine divided by three, nine: how many three's? Peter, nine: how many groups of three? Three! Yes, one, two, three; four, five, six; seven, eight, nine. . . . If you've had a number like this, because you've known your tables, you've been able to know the answer. . . . Thirty-six: How many three's? Twelve. Good boy. With four times. Eight: how many four's? Divided by four is two. . . .

Now we come to a point where the numbers get a lot harder. If we want to know 666: how many lots of three's? We do it in a different way. . . . And, in order that it makes sense and you know where you are going, this is the way you set it out. . . .

Say, the word 'division'. They are called division sums. Division sums. . . . Now these ones are easy. I'll read it to you. Then I want you to read it to me. Three: how many or divided into how many three's are there in 666? How many three's in 666? We start at the beginning. How many three's in six? Hands up, please. . . . How many three's are there in six? Two. How many three's in six? Two. How many three's in six? Two. So that gives you the answer. There are 222 three's in 666. . . .

Despite the considerable emphasis in Victorian and Australian schools on school-based or teacher-developed programs and curricula, on student-centred methods and on realistic problem-solving settings, lessons characterised by a large amount of teacher talk and students having a largely passive role continue to be prevalent.

There is a striking similarity between the content and structure of both of these lessons, taught by experienced, thoughtful and caring teachers to groups of students in suburban schools in Melbourne, Australia, and those described by Romberg (1988) and Leinhardt and Putnam (1987) for the American school context.

Contemporary researchers have studied the consequences for students of learning mathematics in classrooms in which the material presented is in carefully planned steps consistent with the teachers' rather than the students' conceptual frameworks:

There is no reason to assume that a child will interpret a given situation in the way that seems 'obvious' to an adult or a teacher, nor can one assume that a child necessarily will 'see' that a particular way of proceeding must necessarily result in the more standard adult's solution. . . . (L)ittle commonality can be assumed between the teacher's and the children's conceptual structures. (Steffe & Cobb, 1988, p. viii)

In-depth interviews with students about their interpretations of teacher explanations during mathematics classes and the applications of these perceived explanations to exercises and problems posed revealed the extent to which students are engaged during lessons, listen to the teacher and struggle to learn from the expositions given (Leder, 1990). Often, however, the interpretations made and the learning achieved are at variance with those intended by the teacher. Examples of students making 'creative' adjustments to teacher explanations to make them consistent with their own conceptual framework abound. Such efforts can result in inappropriate 'carrying' practices, equating division with subtraction, or using partially correct algorithms. We also identified groups of students who seemed discouraged by their attempts to reconcile with their own conceptual framework the explanations and sequencing thought to be most appropriate and logical by the teacher and resorted to rote recall.

FUTURE DIRECTIONS

The work discussed in the first half of this chapter used relatively large samples and statistical techniques to draw inferences about gender differences in mathematics learning. By focussing on group differences rather than similarities, this approach tends to highlight, reinforce and perpetuate popular beliefs and stereotypes about gender differences in mathematics learning. In the alternative descriptions of mathematics lessons, on the other hand, there was a greater emphasis on gathering information at a micro level (i.e., about the ways in which individual students learn mathematics, by listening extensively to them talking about their work). In fact, Leder and Fennema (1993) argued that intensive explorations with much smaller numbers of students are a timely addition to work on gender differences:

> Not until we somehow get into the 'black box' of the learner's mind will we know whether many of the things assumed to be detrimental or helpful are indeed influences on a learner. For example, is it really detrimental to females that the teacher spends more time with males? Or is the major detriment a single critical incident involving overt sexism? Not until we know what the learner is thinking can we answer such questions. (Leder & Fennema, 1993, pp. 198–199)

There is considerable overlap between the above research methodology and that advocated by a growing number of mathematics educators working within the constructivist paradigm, which emphasises the influence of the mix of personally-constructed conceptions, attitudes, perceptions and abilities on the learning of formal mathematics. Although the role of teachers still is considered to be crucial within this broad constructivist framework, teachers are seen as facilitators of the learning process, not as givers of information to be assimilated passively by the learner. To achieve this, the student's rather than the teacher's cognitive framework should shape the development of the lesson. Teachers should provide opportunities for students to try out and discuss their mathematical ideas and to test them against the understandings of their peers. Furthermore, the learning environment should be supportive so that embryonic knowledge is not jeopardised by ridicule or the insensitivity of others. In short, the classroom climate should be sufficiently flexible to accommodate the full range of student needs. In this way, all students, both females and males, should learn mathematics in a rich and nurturant environment. Further work is needed to increase our understanding of the multifaceted interaction between teacher knowledge, classroom practices and student learning. Undoubtedly, however, teachers will need considerable support – both organisational and content-related – if the goal of true equity in the mathematics classroom is to be achieved.

La Trobe University, Melbourne, Australia

REFERENCES

Australian Council for Educational Research (1977). *Operations test*, Melbourne, Australian Council for Educational Research.

Brophy, J. & Good, T. (1970). 'Teacher-child dyadic interaction: A manual for coding classroom behavior', in A. Simon and E. Boyer (eds.), *Mirrors for behavior: An anthology of classroom observation instruments* (1970 supplement Vol. A), Philadelphia, PA, Research for Better Schools, 83.2-1–83.2-103.

Eccles, J.S. & Blumenfeld, P. (1985). 'Classroom experiences and student gender: Are there differences and do they matter?', in L.C. Wilkinson and C.B. Marrett (eds.), *Gender influences in classroom interaction*, New York, Academic Press, 79–114.

Grimshaw, P. (1982). 'Introduction', in P. Grimshaw and L. Strahan (eds.), *The half open door*, Sydney, Marle and Iremonger, 1–9.

Harris, M.J. & Rosenthal, R. (1985). 'Mediation of interpersonal expectancy effects: 31 meta analyses', *Psychological Bulletin* (97), 363–386.

Leder, G.C. (1986, April). *Gender linked differences in mathematics learning: Further explorations*, paper presented at the Research Presession to the 64th annual meeting of the National Council of Teachers of Mathematics, Washington, DC.

Leder, G.C. (1987). 'Teacher-student interaction: A case study', *Educational Studies in Mathematics* (18), 255–271.

Leder, G.G. (1988). 'Teacher-student interactions: The mathematics classroom', *Unicorn* (14), 161–166.

Leder, G.C. (1989). 'Gender differences in mathematics learning revisited', in G. Vergnand, J. Rogalski and M. Astique (eds.), Proceedings of the 13th International Conference of Psychology of Mathematics Education (Vol. 2), Paris, GRD, dactique CNRS, 218–225.

Leder, G.C. (1992). 'Mathematics and gender: Changing perspectives', in D.A. Grouws (ed.), *Handbook of research on mathematics teaching and learning*, New York, Macmillan, 597–622.

Leder, G.C. (1993). 'Teacher-student interactions, mathematics, and gender', in E. Fennema and G.C. Leder (eds.), *Mathematics and gender*, Brisbane, University of Queensland Press, 149–168.

Leder, G.C. & Fennema, E. (1993). 'Gender differences in mathematics: A synthesis', in E. Fennema and G.C. Leder (eds.), *Mathematics and gender*, Brisbane, University of Queensland Press, 188–199.

Leder, G.C. & Gunstone, R.F. (1990). 'Perspectives on mathematics learning', *International Journal of Educational Research* (14), 105–120.

Leder, G.C., Taylor, P.J. & Fullarton, S. (1994). 'Australian mathematics competition questionnaire', work in progress, La Trobe University.

Leinhardt, G. & Putnam, R.T. (1987). 'The skill of learning from classroom lessons', *American Educational Research Journal* (24), 557–587.

Romberg, T.A. (1988 November). *Principles for an elementary mathematics program for the 1990s*, Revised paper prepared for the California Invitational Symposium in Elementary Mathematics Education, San Francisco, CA.

Rowe, M.B. (1974a). 'Wait-time and rewards as instructional variables that influence language, logic, and fate control, part one: Wait-time', *Journal of Research in Science Teaching* (11), 81–94.

Rowe, M.B. (1974b). 'Relation of wait-time and rewards to the development of language, topic, and fate control, part II: Rewards', *Journal of Research in Science Teaching* (11), 291–308.

Sampson, S.N. (1989). 'Introduction', in G.C. Leder and S.N. Sampson (eds.), *Educating girls: Practice and research*, Sydney, Allen and Unwin, 1–11.

Steffe, L.P. & Cobb, P. (1988). *Construction of arithmetical meanings and strategies,* New York, Springer-Verlag.

PATRICIA F. MURPHY

9. ASSESSMENT PRACTICES AND GENDER IN SCIENCE

In recent years, growing awareness of the relationship between assessment and learning has resulted in several major critiques of existing practice and proposals for reforms at national and regional levels. For example, in the USA, the almost exclusive use of paper-and-pencil multiple-choice tests has been challenged by numerous bodies and the need for new and more varied assessment methods has been emphasised (Murname & Raizen, 1988; Raizen et al., 1989). Central to this new approach is the use of carefully constructed performance tasks that give students opportunities to demonstrate and apply their understanding as they would 'in the world outside of school' (Marzano et al., 1994).

In the UK, after a decade of research into alternative assessment practices (Broadfoot et al., 1988; Brown, 1989; DES, 1989a), the national assessment system advocated for England and Wales (DES, 1988a) involved the use of broadly-based tests which employ a 'wide range of modes of presentation, operation and response' and the combination of results of these tests with school-based teacher-made assessments. In New Zealand, the working party on Assessment for Better Learning (DOE, 1989) made similar recommendations, although the approach to the administration of the tests and the use of the results differed.

These few examples of national initiatives reflect an emerging consensus that improved assessment practice will lead to improved learning. They reveal also a widespread concern about the validity and interpretability of assessment findings. The sources of assessment invalidity generally referred to in the literature include:
- the mismatch between assessment methods and the nature of subjects (particularly in science);
- the narrowness of the definition of students' achievements;
- the bias inherent in certain assessment procedures;
- the manner of describing achievement – in relative terms, as opposed to absolute;
- the manner of eliciting responses – in time-constrained, paper-and- pencil mode only;
- the isolation of the students from the assessment process.

In response to these concerns, educational assessment has focussed on the *quality* of educational achievement. Consequently, it is ipsative rather than normative and is based on criterion-referenced views of achievement. Criterion-referenced assessment measures students' performance against explicit criteria and can serve both *summative and formative* purposes. Summative assessments attempt to provide a measure of the current overall achievement of a student. Formative assessments focus on the *nature* of students' achievements and needs, and thus are able to inform

105

L.H. Parker et al. (eds.) Gender, Science and Mathematics, 105–117.
© 1996 *Kluwer Academic Publishers. Printed in the Netherlands.*

the planning of their future learning.

Assessment can benefit the education process only if an individual's achievements can be assessed fairly. One limiting factor is the potential bias in assessment
items and procedures, a problem which is recognised widely, but not understood
well. The question of whether a test is biased is extremely difficult to answer. The
validity of assessment depends on the justice of the interpretations made of test or
item scores (Messick, 1989). Valid interpretation depends crucially on assessors understanding the significant characteristics of learners and assessment tasks and how
these interact. Assessors have to establish not just what students know, but how they
come to understand things in certain ways. An emphasis on formative assessment
could help the situation of girls' alienation from, and possible underachievement in,
science by revealing the effects of gender differences and indicating ways to address
them. To explore the implications of gender differences for assessment in science,
this chapter looks at some of the relevant international research in the area and examines the findings in the light of the results of science surveys carried out in the UK.

GENDER DIFFERENCES IN ACHIEVEMENT –
THE INTERNATIONAL SCENE PAST AND PRESENT

Much of the literature critical of present assessment practice argues not only for alternatives to multiple-choice tests, but also for the use of a variety of methods. The UK
Assessment of Performance Unit's (APU) science project provides a model for such
an alternative assessment in science (DES, 1989a). The APU conducted national
surveys of 11-, 13- and 15-year-old students in England, Wales and Northern Ireland
annually from 1980–1984. The APU science surveys assessed both the skills and
content of science. Three of the tests were practical and only a small number of the
test questions used were multiple-choice (Murphy & Gott, 1984). In addition, questionnaires were used to collect data on issues thought to influence science performance, such as out-of-school activities and interests (DES, 1988b, 1989b).

Table I (Murphy, 1991) includes some of the overall test results for gender differences for the five APU surveys. The table also includes the results of three other
surveys, namely, those of the International Association for the Evaluation of Educational Achievement (IEA) (Comber & Keeves, 1973), the USA National Assessment
of Educational Progress (NAEP, 1978) and the British Columbia Science Surveys
(BCSS) (Hobbs et al., 1979). In all three surveys, the tests were not practically based
and the questions were largely multiple-choice.

The BCSS results show boys ahead of girls only in physics and measurement
skills. The IEA and NAEP surveys, however, found that boys outperformed girls
across the tests, even when their curriculum backgrounds were similar. This was not
the case in the APU surveys. When students with the *same* curriculum backgrounds
were compared, all gender gaps in performance disappeared at age 15 years, except
those for the sections *Applying Physics Concepts* (written test) and *Making and Interpreting Observations* (practical test). Moreover, the gap in physics achievement

TABLE I
Some international survey results

Areas tested	APU results	Results of other surveys		
		IEA	NAEP	BCSS
Use of graphs tables and charts	$B_{15} > G_{15}$			B = G
Use of apparatus and measuring instruments	$B_{15} > G_{15}$	B > G		B > G
Observation	G > B			
Interpretation	$B_{13,15} > G_{13,15}$		B = G	
Application of:				
Biology concepts	B = G	B > G	B > G	B = G
Physics concepts	B > G	B > G	B > G	B > G
Chemistry concepts	$B_{15} > G_{15}$	B > G	B > G	B = G
Planning investigations	B = G	B > G		B = G
Performing investigations	B = G			

In some cases there is a suffix to indicate the age at which the gender difference emerges. Where there is no suffix, this indicates that the difference or similarity in scores occurred across the ages tested.

between girls and boys established at age 11 increased with age (Johnson & Murphy, 1986), a finding reported in the other surveys.

The APU questionnaire results showed, across all ages, girls' interests lying in biological/medical applications and boys' interests involving physics and technological applications. The same polarisation was evident in the pastimes and hobbies reported by the students. Girls at age 13 and 15 years, unlike boys, saw science as having little relevance to their potential careers and, at age 15, more girls than boys considered physics and chemistry to be difficult (DES, 1988b, 1989b).

More recent surveys by the NAEP continue to show a consistent male advantage in science (Mullis *et al.*, 1990). In addition, two international studies using NAEP items (again, with no practical element) have been carried out as part of the International Assessment of Mathematics and Science (IAEP). The first survey, in 1988, assessed only 13-year-olds and found boys outperforming girls significantly in all populations except those of the UK and the USA (Lapointe *et al.*, 1989). The second survey, in 1990, included 14 countries at age 9 years and 19 countries at age 13 years. Boys once more tended to do better than girls in nearly all the countries participating, more so at age 13 than at age 9 and particularly in physical and earth and space sciences. In a number of countries, girls were ahead of boys on questions about the 'nature of science' (Foxman, 1992). Once again, there were no statistically significant gender differences for the English sample at age 9 or 13. The students' responses to the statement 'Science is for boys and girls about equally' revealed that a high proportion of UK students of both ages agreed (86% at age 9 and 97% at age 13).

The more recent British Columbia assessments in science included practical test items similar to those used by the APU. The 1991 report on students in grades 4, 7 and 10 showed, in the written tests, females out-performing males on:

... many of the process skills, and on items involving conceptual planning, abstract and critical thinking and the nature of science and safety. Males perform higher on items involving measurement, lab techniques, numerical problem solving, earth/space science and energy topics, particularly electricity. (Bateson *et al.*, 1991, p. 18)

The differences measured, however, were significantly less than those measured previously. Males outperformed females on practical items using scientific equipment but, on the practical environmental items, the gender gap was in favour of females and increased with age. In addition, the attitude measures revealed similarities between males and females, with the exception that 'females are more strongly positive than males in the area of environmental issues. These differences tend to become larger as students get older' (Bateson *et al.*, 1991, p. 8).

The various survey results suggest some interesting links between male and female performance, attitudes, curriculum experience and the forms of assessment used. A study in Thailand (Harding *et al.*, 1988) involving six schools and 1,481 students provides further support for this. The tests used included practical and written forms covering a broad range of science content and skills. The results showed girls at age 16–18 years performing at least as well as boys in physics and better than boys in chemistry. In laboratory tests, girls outperformed boys in both physics and chemistry. The researchers suggest several reasons for these results. In Thailand, the teaching approach is practical and chemical tasks are said to have a 'feminine' image, as do some of the practical physics tasks. In England, where there is no overall difference between males and females, the science curriculum is practically orientated with a focus on investigations and problem-solving. Recent research has shown that girls following problem-solving science courses in the UK report a strong liking for science (Bentley & Drobinski, 1995).

In summary, the research suggests that girls' attitudes to, and achievement in, physics specifically, and science generally, are related to the cultural expectations of girls and boys and how these are reflected in the organisation and values underpinning the school curriculum and its assessment.

FACTORS WHICH LEAD TO INVALIDITY IN ASSESSMENT OUTCOMES

In their summary of sex role development research, Wilder and Powell (1989) comment on the different ways in which parents respond to boys and girls and encourage them to interact with their environment and with other people. Parents expect different behaviours from boys and girls and these expectations are reflected in the activities, toys and home environments which they provide for them and in the different types of feedback they typically mete out. A consequence of this is that children develop different ways of responding to the world and making sense of it, and these ways influence how children learn and what they learn.

In similar ways, boys and girls experience schooling differently. For example, the interactions between teachers and boys and teachers and girls have been found to vary in frequency, duration and content. Consequently, boys and girls develop different perceptions of their abilities and relationships with academic disciplines. Teachers' expectations and judgements of girls' and boys' achievements and needs also have been found to vary in stereotypical ways. Children's judgements closely reflect those of their teachers and are again domain dependent. The consequences of this for assessment are considerable.

Differences in Experiences and Perceptions of Purpose

There is much evidence about students' differential experiences out of school and some of the direct effects that these have on measured achievement in school. For example, both the NAEP and the APU surveys found that boys' and girls' experience of scientific equipment and apparatus out of school differed. Boys were more likely to have had experience of a greater range of relevant equipment and instruments than girls. Where gender differences arose in the APU surveys of the use of measuring instruments, they were in favour of boys and on precisely those instruments with which the students had reported more experience out-of-school (DES, 1988b, 1989b). These performance differences increased in range and magnitude as students progressed through school (Johnson & Murphy, 1986). The NAEP and APU surveys also revealed gendered patterns of participation in out-of-school activities and hobbies. In relation to science, these were seen to provide boys with greater opportunities than girls to accumulate a working knowledge of mechanics, to develop spatial abilities and to gain an early familiarity with electricity.

If the results of assessments are to be interpreted correctly, it is important to collect the kind of information presented above. The different experiences of students affect not only the skills and knowledge which they develop, but also their understanding of the situations and problems in which to apply them. For example, boys in the APU surveys were better able to use ammeters and voltmeters, yet they did not report using these outside of school. Boys, however, continue to play more than girls with electrical toys and gadgets. Such play allows them to become familiar with the effects of electricity and to develop a tacit understanding of how it can be controlled.

Another example of the influence of different pastimes relates to girls' well-documented lack of experience of tinkering and modelling activities. The majority of UK primary schools provide children with ample access to modelling resources, such as LEGO. Teachers, however, often are discouraged by girls' apparent lack of interest in modelling activities and the girls' tendency to replicate only domestic situations. Boys *choose* to make models out-of-school and their play allows them to establish the link between their purposes and the potential of the modelling medium. It is this relationship that girls need opportunities to explore. To facilitate this, teachers have to work with girls to help them to identify the problems which they find motivating and for which modelling serves a useful function. The same holds true for other areas of students' inexperience.

In summary, the purpose for carrying out various tasks becomes a critical issue, especially for assessment. Typically, in assessment tasks, purpose is either missing or assumed to be implicit and unproblematic, with the result that students have to import their own sense of purpose in order to make sense of tasks. The purpose defines what knowledge students consider to be appropriate to draw on and, ultimately, what task they tackle. The APU results indicate that, relative to boys, girls' purposes are more often at odds with those of science assessors and their performance more frequently misinterpreted as a consequence.

Differences in Ways of Experiencing

The results of the APU science surveys show that girls and boys react to the same item content differently, irrespective of what is being assessed. Questions which involve content relating to health, reproduction, nutrition and domestic situations generally are answered by more girls than boys. The girls also tend to achieve higher scores on these questions. In situations with a more overtly 'masculine' content (e.g., building sites, submarines, cars or anything with an electrical content), the converse is true. For example, girls and boys at ages 11 and 13 years are equally competent at interpreting data in tabular form. Yet, if the table is about the different flowering stages of a plant, then some boys do not respond, whereas girls tackle the item with confidence and overall obtain a higher score. On the other hand, if the table is about an extract from a spare parts manual, the reverse happens. For some students, the items are not about interpreting information but about their experience of the content. Assessors unwittingly can skew overall test results by the selection of content. This has been noted in the NAEP surveys (Gipps & Murphy, 1994) and in mathematics and English assessments in the UK (Stobart et al., 1992).

The dilemma for assessors is that such items are sources of assessment invalidity, because failure on them does not arise because students lack the knowledge or skill being assessed. But, significantly, such items can provide teachers with invaluable formative insights to aid curriculum planning, when it is remembered that students' alienation from certain activities and content related to science can lead ultimately to underachievement, because they fail to engage with certain learning opportunities.

Girls' and boys' different ways of experiencing the world impact also on their predictions of their own success on the practical observation items. The results of the APU science surveys showed that girls underestimated their performance in this regard, whereas boys generally overestimated theirs. Girls' lower expectations of success, coupled with their unfamiliarity with certain science-related experiences, means that they approach some learning situations with diffidence and fear (Randall, 1987). In the APU surveys, more girls than boys rejected investigations for which they were expected to use overtly scientific equipment (a finding corroborated by the 1991 British Columbia survey). However, if the same students were given the identical investigations set in everyday contexts with everyday equipment, more girls, and indeed more boys, felt able to tackle them and to provide everyday solutions (DES, 1989b). In their approach to such investigations, however, students tended not to control variables or collect quantified data, a finding replicated in recent research (Foulds *et al.*, 1992). A gender-inclusive science assessment task, therefore, would have to be set in an everyday context and to involve a problem that can be solved only by using some degree of rigour and accurate scientific equipment.

In summary, it is clear that the effects of alienation must be distinguished from differences in ability, in order to interpret achievement correctly. In most instances, the effects of alienation result in the assessment of factors unrelated to the construct identified by the assessor.

Differences in Problem Perception

The ways in which girls and boys perceive assessment tasks or problems are often very different, given the same circumstances. Girls tend to value the circumstances in which assessment tasks are set and take account of them when constructing meaning in the task. They do not abstract issues from their context. Conversely, as a group, boys tend to consider issues in isolation and judge the content and context to be irrelevant. This latter approach generally is assumed to be the norm in both assessment and classroom practice in science.

In the APU studies, an example of this difference in problem perception occurred when children in primary schools were asked to design model boats to go around the world. Part of their task involved investigating how much load their boats would support. Some of the girls were observed collecting watering cans, spoons and hair dryers. The teacher interpreted their action as a *failure* to understand the problem. However, as the girls explained, to go around the world one needs to consider the boat's stability in monsoons, whirlpools and gales. The teacher had moved on from the context which she had introduced to create a sense of purpose for the task. For the girls, the context remained an integral part of the task.

In another situation, students in secondary schools were investigating which of two materials would keep them warmer as a jacket when stranded on a cold, windy mountainside. To find this out, they were expected to compare how well the materials kept cans of hot water warm. Again girls were seen to be doing things that teacher assessors judged to be 'off task'. They cut out prototype jackets, dipped the materials in water, and blew cold air through them. These girls took seriously the human dilemma presented. It mattered how porous the material was to wind, how waterproof it was and whether it was suitable for making a jacket. Current research into primary science learning reported by Murphy *et al.*, (1994) has noted similar effects in mixed groups tackling science investigations.

Other instances of students' different perceptions of problems reflect the ways in which girls are encouraged, outside of school, to be concerned with everyday human needs. The boats which the students designed in the task mentioned earlier covered a wide range, but there were striking differences between boys and girls. Typical examples are shown in Figure 1. Boys' boats typically were power boats or battleships. The detail included varied, but generally there was elaborate weaponry

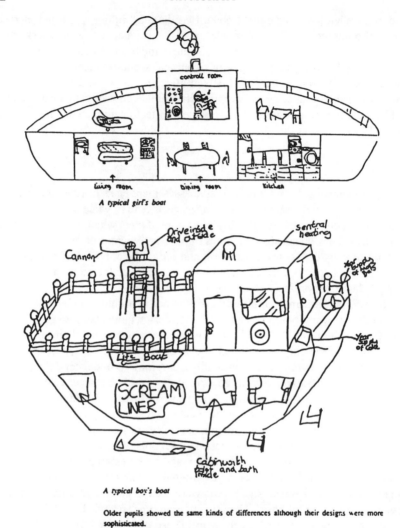

Figure 1.   Examples of childrens' designs of model boats to go around the world.

and next to no living facilities. Other features included detailed mechanisms for movement and navigation. The girls' boats were generally cruisers, with a total absence of weaponry and a great deal of detail about living quarters and requirements, including food supplies and cleaning materials (notably absent from the boys' designs). Very few of the girls' designs included any mechanistic details.

Sorensen's research (1990), which looked at grade 7 children in Denmark when they built model houses and installed electricity in them, yielded findings similar to those cited above. The girls' houses were decorated and made to look real, with much thought given to the appropriate location for light switches. The boys' houses, were generally LEGO block models with correct circuitry but sometimes with light switches on the outside of the house. For the boys, the context of the 'home' was viewed as irrelevant to the school-task. Thus, they did not pay attention to details like the location of switches.

The British Columbia survey (Bateson *et al.*, 1991) found gender differences in performance on socioscientific issues, and these also are relevant to this discussion. Females, more than males, identified different points of view and inclined towards caring and conserving view points. Males favoured more utilitarian perspectives.

Girls' concern with decor, diet and hygiene generally is dismissed as symbolic of a particular learning style:

Girls' tendencies to be distracted by powerful cues or true but irrelevant facts seem to reflect the 'hesitant, dependent, anxious, unmotivated, help searcher learner'. (Levin *et al.*, 1987, p. 111)

In a typical assessment situation, these examples of 'girls'' solutions are likely to be judged as inappropriate, either because 'girls'' problems are not recognised or because they are not valued by assessors. There are two issues to consider here. First, on what grounds are girls', and other groups of students' views of relevance accorded low status in assessments? Second, if students' views are to be denied, then interpretations of assessment results should indicate that students, in reality, have addressed alternative tasks, not that they have failed to provide 'correct' responses to the assessor's tasks.

Differences in Styles of Expression

Another potential source of assessment bias arises from the different reading interests that girls and boys typically develop. Girls enjoy a wide range of books, whereas boys tend to read non-fiction, particularly technical and hobby manuals. These preferences affect the styles of writing which students adopt. Girls choose to communicate their feelings about phenomena in extended and reflective composition, while boys provide episodic, factual, commentative detail (Gorman *et al.*, 1988). The significance for assessment is that differences in style alter assessors' perceptions of the content of students' responses.

In science, a typical assessor has expectations of an appropriate response format. Hence, in response to an observation task, it is expected that a list will be drawn up, categorised or ordered with respect to the type of observations made. Figure 2 displays an extract of a typical 'girls'' style of response which would not be expected normally. The task set was to write a description comparing two pictures of demoiselle flies, the aim being '*to emphasise the differences most important for telling them apart*'. The response reflects the style of writing that girls commonly meet in the literature that they choose to read.

Style contributes to the stereotyped images that teachers have about what areas of the curriculum are appropriate for which gender. The study of Goddard Spear (1987)

It is one of those lazy hot days in summer when everything is warm and very quiet. The trees surrounding the lake at the bottom of the hill are swaying silently and the ripples on the lake give the impression of peace and tranquillity.

At the end of the lake are reeds and lilies. Flies buzz dozily among the tall grasses, looking for food. Bees laze among the pollen filled lilies, drinking their sweet nectar and the demoiselle flies perch motionless on the tall green fronds of the reeds.

There are two in particular, one male, one female, that catch my eye as I lie against the sturdy trunk of an ancient oak. They are the most beautiful creatures I have ever seen, but they are both different.

One has lacy wings, so clear I can see the water's edge through them. Its colouring is of brilliant pinks and blues, and it stands out amongst the yellow buttercups that surround it. Its abdomen is long, like a finger, and incredibly thin. It looks so fragile, as though any sudden movement may snap it, like a twig. Its lacy wings stretch back, almost to the full length of the abdomen, like delicate fans, cooling its body. Its head is small but bold. It is completely blue with piercing black eyes on either side of its head. The legs of this magnificent creature are long and black, with what look like hairs of the finest thread, placed at even spaces down each side.

The thorax, the part next to its head, is large. It is not as slender as the abdomen, but it is very sleek, with patches of blue and black reflecting the brilliant sunlight.

As I watch, its head rotates and then suddenly it has disappeared, hovering over the lake.

The other demoiselle fly still remains. This is not such a beautiful creature as the first, but it has striking markings. The wings are a dull brown in colour. They are much wider and not as long. They appear to be much more powerful than the lacy, delicate wings of the other fly. The abdomen of this creature is much thicker. It is dull brown, like the wings, but has flecks of mauve and grey. It appears, just as with the wings, to be much stronger, more substantial, and more useful.

Figure 2.   A girl's description of the differences between pictures of two demoiselle flies.

showed that the same piece of science writing, when attributed to girls, received lower marks from teachers than when it was attributed to boys. Thus, the increased use of students' self-assessments and student-selected pieces of work will not enhance the validity of assessment if assessors do not distinguish between features of style and the criteria that they are assessing.

DISCUSSION: GENDER FAIR ASSESSMENT

How might we tackle the idea of gender-fair assessment? Rennie (1987), reporting a study of 13-year-olds in Western Australia, found that the relative inexperience of girls can be overcome in an activity-orientated style of science teaching. Similarly, the science curriculum of Thailand, which appears to support girls' learning, is described as activity-based, learner-centred and focussed on novel aspects of inquiry and discovery. The implementation of a broadly based, investigative science curriculum in England has coincided with an overall increase in girls' achievements in science at age 14 years (CATS, 1991). There also is recent evidence that a problem-solving approach to science curriculum is increasing girls' liking for science in the UK (Bentley & Drobinski, 1995). In relation to assessment, it appears that the same effect exists. The APU observation tasks, on which girls across the ages achieved higher scores than boys, were novel and activity-based. These kinds of tasks show few of the content effects described earlier, apparently because students perceive the tasks in terms of the observations required, or alternatively because the tasks allow students to formulate their own hypotheses and to test them out before reaching a conclusion. Interestingly, these tasks tend also to be the tasks that girls and boys prefer. Thus, the combination of an active response and an open approach makes these tasks accessible to both girls and boys.

One assumed strategy for achieving gender-fair assessment is to eradicate bias by using items for which girls and boys achieve equivalent scores. However, gender bias arises from a complex interaction of effects related to features of assessment tasks. For example, a multiple-choice question involving a particular biological content, focussing on distinctions between colours, might well show girls and boys as groups achieving equal scores. However, some boys will fail to respond because the content alienates them, while others might have difficulties because they typically do not consider colours as relevant data. On the other hand, some girls will not respond because they cannot see the one 'right' answer that they are supposed to. If results are not interpreted at this level of detail, the information that they provide might misinform curriculum planning, because all students who fail to provide the 'answer' are assumed not to have the knowledge being assessed. A further problem with attempting to eradicate bias from assessment tasks is that not enough is known about the features of tasks which affect students' ability to construct meaning in them.

An alternative approach would be to provide tasks that allow all students to express their interests and understandings in a manner that suits them. However, in tasks of this kind, students will see different problems and thus find different solutions. The problem for assessment is that one cannot compare such responses – they are just different. Nor can one assume that a task of this kind will provide information that 'fits' a subject criterion neatly. Boys' and girls' problems draw on, and reflect, different aspects of their scientific understanding. To deal with assessment outcomes of this type, it is necessary to move away from simplistic notions of a scoring scheme targeting a 'right' answer, to a consideration of students' strategies in the light of the problems which they perceive. Such assessments provide teachers with the information they need about an individual's understanding. Teachers can act on the insights provided to enable individuals to develop further their scientific understanding in a way that is integrated with their existing knowledge of the world and themselves. Such knowledge is robust rather than fragmented, and can be used by students to make sense of new situations. Insights from students' responses also provide a rich

array of alternative views of the world which students can share. These experiences
enable them to develop an awareness of the potential for choice in the selection and
use of their knowledge.

Open University, Milton Keynes, UK

REFERENCES

Bateson, D., Erickson, G., Gaskell, J. & Wideen, M. (1991). *Science in British Columbia*, British
   Columbia, Ministry of Education.
Bentley, D. & Drobinski, S. (1995). 'Girls, learning and science in the framework of the National
   Curriculum', *Curriculum Journal* 6(1), 79–100.
Broadfoot, P.M., James, M.E., McMeeking, S., Nuttall, D.L. & Stierer, B.M. (1988). *Records of achieve-
   ment: Report of the national evaluation of pilot schemes*, London, Her Majesty's Stationery Office.
Brown, M. (1989). 'Graded assessment projects: Similarities and differences', in P. Murphy and B.
   Moon (eds.), *Developments in learning and assessment*, London, Hodder and Stoughton, 300–311.
Comber, L.C. & Keeves, J.P. (1973). *Science education in nineteen countries*, Stockholm, Almqvist
   and Wiksell.
Consortium for Assessment & Testing in Schools (CATS) (1991). *Pilot report 1991: Key stage 3 sci-
   ence*, London, Schools Examination and Assessment Council.
Department of Education (DOE) (1989). *Assessment for better learning – a public discussion docu-
   ment*, Wellington, New Zealand, Department of Education.
Department of Education & Science (DES) (1988a). *Task Group on Assessment and Testing: A report*,
   London, DES.
Department of Education & Science (DES) (1988b). *Science in schools age 15: Review report*, Lon-
   don, Her Majesty's Stationery Office.
Department of Education & Science (DES) (1989a). *National assessment: The APU science approach*,
   London, Her Majesty's Stationery Office.
Department of Education & Science (DES) (1989b). *Science in schools age 13: Review report*, Lon-
   don, Her Majesty's Stationery Office.
Foulds, K., Gott, R. & Feasey, R. (1992). *Investigative work in science*, Durham, The University of
   Durham.
Foxman, D. (1992). *Learning mathematics and science: The second IAEP in England*, Windsor, UK,
   National Foundation for Educational Research.
Gipps, C. & Murphy, P. (1994). *A fair test? Assessment, achievement and equity*, Milton Keynes, Open
   University Press.
Goddard-Spear, M. (1987). 'The biasing influence of pupil sex in a science marking exercise', in A.
   Kelly (ed.), *Science for girls?* Milton Keynes, Open University Press, 46–51.
Gorman, T.P., White, J., Brook, G., Maclure, M. & Kispal, A. (1988). *Language performance in schools:
   Review of APU language monitoring 1979–1983*, London, Her Majesty's Stationery Office.
Harding, J., Hildebrand, G. & Klainin, S. (1988). 'International concerns in gender and science/tech-
   nology', *Educational Review* (40), 185–193.
Hobbs, E.D., Boldt, W.B., Erickson, G., Quelch, T.P. & Sieban, B.A. (1979). *British Columbia science
   assessment 1978, general report 1*, British Columbia, Ministry of Education.
Johnson, S. & Murphy, P. (1986). *Girls and physics*, London, Department of Education and Science.
Lapointe, A., Mead, N. & Phillips, G. (1989). *A world of differences: An international assessment of
   mathematics and science*, Princeton, NJ, Educational Testing Service.
Levin, T., Stobar, N. & Libman, Z. (1987). 'Girls' understanding of science: A problem of cognitive or
   affective readiness', in J. Daniels and J.B. Kahle (eds.), *Contributions to the Fourth GASAT Conference*, Vol. II, Ann Arbor, MI, University of Michigan, 104–112.

Marzano, R.J., Pikering, D & McTighe, J. (1994). *Assessing student outcomes*, Association for Supervision and Curriculum Development, Alexandria, VA.

Messick, S. (1989). 'Meaning and values in text validation: The science and ethics of assessment', *Educational Researcher* 18(2), 5–11.

Mullis, I.V.S., Owen, E.H. & Phillips, G.W. (1990). *Accelerating academic achievement: A summary of findings from 20 Years of NAEP*, Princeton, NJ, Educational Testing Service.

Murname, R.J. & Raizen, S.A. (eds.) (1988). *Improving indicators for the quality of science and mathematics education in grades K–12*, Washington, DC. National Academy Press.

Murphy, P. (1991). 'Gender and practical work' in B. Woolnough (ed.), *Practical work in science*, Milton Keynes, Open University Press, 112–122.

Murphy, P. & Gott, R. (1984). *Science assessment framework age 13 and 15: Science report for teachers: 2*, Hatfield, Association for Science Education.

Murphy, P., Scanlon, E., Hodgson, B. & Whitelegg, E. (1994). 'Developing investigative learning in science – the role of collaboration', in *Proceedings of the European Conference on Curriculum*, Enschede, The Netherlands, University of Twente.

National Assessment of Education Progress (NAEP) (1978). *Science achievement in schools: A summary of results from the 1976–1977 national assessment of science*, Washington, DC, Education Commission of the States.

Raizen, S.A., Baron, J.G., Champagne, A.B., Haertel, E., Mullis, I.N.V. & Oakes, J. (1989). *Assessment in elementary school science education*, Washington, DC, National Center for Improving Science Education.

Randall, G.J. (1987). 'Gender differences in pupil-teacher interactions in workshops and laboratories', in G. Weiner and M. Arnot (eds.), *Gender under scrutiny*, Milton Keynes, Open University Press.

Rennie, L.J. (1987). 'Out of school science: Are gender differences related to subsequent attitudes and achievements in science?', in J. Daniels and J.B. Kahle (eds.), *Contributions to the Fourth GASAT Conference*, Vol. II, Ann Arbor, MI, University of Michigan, 9–17.

Sorensen, H. (1990). 'When girls do physics', in I. Granstam and I. Frostfeldt (eds.), *Contributions to GASAT Conference*, Jönköping, Sweden, Jönköping University College, 42–52.

Stobart, G., White, J., Elwood, J., Hayden, M. & Mason K. (1992). *Differential performance at 16+: English and mathematics*, London, Schools Examination and Assessment Council.

White, J. (1988). *The language of science*, London, Her Majesty's Stationery Office.

Wilder, G.Z. & Powell, K. (1989). *Sex differences in test performance: A survey of the literature*, College Board Report No. 88–1, New York, College Entrance Examination Board.

10. GENDER EQUITY AND THE ENACTED SCIENCE CURRICULUM

At the beginning of the 20th century, the world was on the brink of technological revolution. Science and technology were partners in a process of change that rapidly would transform the way in which people lived their lives and shaped their dreams. Today, as we approach the onset of the 21st century, the world is transformed radically from that world of 100 years ago. The computer has found its way into every imaginable facet of human life and has enabled an information revolution to accelerate its way through the latter years of the present decade. This revolution has the capacity to transform the way in which science education proceeds. In the privacy of the home, potential learners can 'surf' cyberspace, capture evocative images of comets colliding with Jupiter, examine data of all forms at an enormous number of sites, communicate with others through the computer on an eye to eye basis, participate in discussions between teachers on how science ought to be taught and learned, and read exchanges on virtually any topic associated with myriad news groups. Although the internet has found its way into only a handful of schools and its use is limited by the availability of computers and networking software, an explosion in the ownership of personal computers has brought the internet into many homes in the USA and around the world. Widespread availability has increased the accessibility of learning resources for science education in a manner that is potentially blind to colour and gender, although not to social class. The advent of the information revolution promises to permeate the conservative boundaries of schooling and science education. It is within this context, one of a possible revolution in the nature of schooling, that it is critical to consider issues of gender equity in science education.

In Western society, women's roles have changed a great deal and equity with men is a goal towards which many institutions aspire. However, science, the engine for so many of the reforms of the 20th century, has not changed to admit women and, in science, there are fewer women than men, particularly in senior positions. Recently, I asked a physicist about the issue of gender in physics. His response provides some insights into the extent of the problem:

The enrolment in our PhD program is only about 10% women, even though we are being fairly aggressive about recruiting them. We even award 'Women in Physics' Fellowships to two or three women in the entering class each year. These women receive an annual stipend of about $16,500 per year, instead of the standard amount of about $14,000 per year. Our gender situation in the PhD program is about the same as it is nationally.

At the undergraduate level, about 30% of the physics majors seem to be women here. My impression is that this is also typical of the national situation. The difference in the proportions of women at the undergraduate and graduate levels is a real problem. It can be argued that part of the problem is that women undergraduates do not have role models among the physics faculty: less than 5% of the tenured

119

L.H. Parker et al. (eds.) Gender, Science and Mathematics, 119–127.
© 1996 *Kluwer Academic Publishers. Printed in the Netherlands.*

physics faculty nationwide are women. In our department of 45 faculty, we went a few years without a single female faculty member. Now we have two, both young. One of them is attacking the shortage of women physicists aggressively; the other would prefer not to worry about this issue.

But, for all of our effort and all of the rhetoric coming out of the national physics community, there are still many instances of heinous and overt sexual harassment in the field. An undergraduate major who did some fantastic research work for me answered the door bell to her apartment one night about three years ago to find one of our faculty members standing there with a bottle of champagne. That faculty member is no longer with the department (although this incident was laughed off by nearly everyone).

The same student ran an overnight shift at the accelerator laboratory with a male graduate student shortly afterward. The male student (who weighed about 200 pounds) made an inappropriate remark about the female student (who weighed about 90 pounds). The female student promptly slugged him. They worked together for the rest of the night and the rest of the year perfectly peacefully. The male student earned his PhD and now has an industrial job as a manager; the female student is now in a physics graduate program.

Another undergraduate female student of mine went to a national laboratory for a summer internship. After about a month, her research adviser there began to proposition her repeatedly. She reported this behaviour to a woman administrator involved in the internship program, who then did absolutely nothing. The administrator had explicit responsibility for addressing sexual harassment complaints. This student earned her Bachelor of Science after a very productive undergraduate research career here, and is now in a Master's degree program in medical physics.

The introductory calculus-based course, which serves mostly engineering students, but also chemistry and physics, has about 40% females. The introductory algebra-based course has about 60% females. Research indicates that women and men in the algebra-based course achieve equally and perceive the learning environment in the same way, so things seem to be OK in that course. No such data are available for the calculus-based course.

Although it is promising to think that 40% of the population of the calculus-based introductory physics enrolments is female, it is a concern that this percentage is not reflected in graduate enrolments. Furthermore, the low proportions of female faculty and graduate students in physics are particularly discouraging, as are reports of harassment against those females who do choose to enter physics. The occurrence of such events is likely to discourage some females from participating in science. Similar signs, that discourage females from participating in science, also might be evident in schools.

PURPOSE

This chapter examines a framework that considers females as minorities in a culture of science and uses research on gender equity from that perspective in order to highlight problems and potential solutions in a manner that differs from the traditional approach.

FEMALES AS MINORITIES IN SCIENCE

Ogbu (1992) provides a framework for considering the education of minorities. Involuntary minorities include people who are in the mainstream culture against their will, due to historical factors such as slavery, conquest, colonisation and forced labour.

Involuntary minorities often experience difficulties with school learning due, in part, to cultural inversion (i.e., a tendency to regard certain forms of behaviour, events, symbols and meanings as inappropriate because of their association with the mainstream culture). In contrast, voluntary minorities are those who have migrated to a country, more or less voluntarily, and desire more economic well being and, in many cases, greater political freedom. They usually experience initial difficulties due to language and culture as well as a lack of understanding of how the educational system works. However, unlike involuntary minorities, voluntary minorities gradually adapt to the mainstream culture and succeed academically.

Members of a group that is regarded as consisting of voluntary minorities can assume the characteristics of involuntary minorities. In terms of the culture of science, it appears as if some females participate minimally, against their will, for the minimum amount of time, and suffer from cultural inversion. They see the culture of science as being inappropriate for them and reject it because they do not fit. Although a female student might construct herself as interested in science and committed to learning with understanding (i.e., as a voluntary minority), she might feel as if she is swimming against a strong current as she endeavours to convince others of her goals. She will feel most comfort when her practices fit the expectations of others in the classroom. When a female walks into a classroom, she is constructed by others in ways that partially reflect her personal characteristics and also those associated with being female. Those in the class, including the teacher, construct themselves and one another. Thus, each female, whether she likes it or not, is constructed in ways that take account of the fact that she is female. Within a given society, each female will feel a pressure to adopt a set of roles that is consistent with the constructions of others and, when others interact with her, the expectations associated with those interactions will be mediated by a realisation that the person with whom they are interacting is female. Over a period of time, these pressures can lead some females to feel unwelcome in the culture and their practices can become more like those of involuntary minorities.

Ogbu argued that minorities have caste-like features: cultural baggage assigned by others with whom they interact. What seems most salient about this phenomenon is that often it is an invisible part of belonging to a culture. Neither the female who has to deal with this cultural baggage nor those who construct females as they do might be aware specifically of what is happening. The construction of self and non-self are a part of the daily routines of individuals in any social setting, a part of normal life within a cultural setting. The normal routines of constructing self and others have been part of the problem of dealing with significant gender differences in science classes. Some teachers do not notice gender differences and, even when they are pointed out to them, they might recognise such differences as the way that things are meant to be.

INVISIBLE CHARACTER OF GENDER DIFFERENCES IN SCIENCE CLASSROOMS

Research in which I have been involved for more than a decade suggests that gender differences in science classes are accepted as a component of what is normal and expected. Indeed, unless teachers and students are presented with evidence to the contrary, they are likely to deny that issues relating to gender equity occur in their classes. For example, at the conclusion of an elementary science lesson on properties of liquids, I spoke to a male teacher about what I had observed. 'Did you know about the gender differences in student responses to your questions?', I asked. 'There were no gender differences in my lesson today', retorted the teacher with a defiant glance towards me. Cautiously I explained that boys had dominated the responses to his questions concerning the properties of liquids and it was not until the end of the lesson, when he introduced questions concerning the addition of detergent to water, that girls participated to any significant extent. On reflection, the teacher grudgingly admitted that the anomalous results that I had described did occur, but he explained that he usually made sure that all students participated to an equal extent.

I had no doubt that this teacher was sincere in his desire to have all students participate to the same extent in classroom interactions. At the same time, I was certain that this particular lesson was not appreciably different from most other science lessons taught by this teacher. The teacher, who was in his early thirties, had been teaching for 10 years and had been recommended to me as an exemplary teacher of grade 5 students. In this lesson, he had taught in a routine manner using strategies that had been honed throughout a decade of teaching. His main conscious thoughts were associated with implementing a curriculum in which his content knowledge was not strong and managing a class that at times was almost out of control. The situation is typical of others we have encountered in a research program that has investigated science education from pre-K to grade 12 (e.g., Tobin & Espinet, 1989; Tobin & Gallagher, 1987; Tobin et al., 1990).

GENDER-RELATED DIFFERENCES IN SCIENCE CLASSES

Although the primary purpose of my research was to understand science teaching and learning better, numerous assertions emerged that were pertinent to gender differences. The findings are based on research undertaken in classrooms in two states of Australia and several states of the USA and are consistent with gender-related differences in performance that have been reported in studies undertaken throughout the English speaking world (Kahle & Meece, 1994). The practices of teachers and male and female students in science classes are such that some females participate actively and can be regarded as voluntary minorities, while the majority of females do not participate to the same extent as males and are not offered the same level of opportunities to learn. These female students reasonably might develop some of the characteristics of involuntary minorities and choose not to participate fully in activities and, when given the choice, opt out of science.

In whole-class settings, we observed males being more involved than females. The nature of the involvement was oral. To a greater extent than females, males either called out a response or reaction or raised their hands as an indication to the teacher that they wanted to respond to a question or contribute to the discourse. Teachers called on students whom they felt could contribute to the flow of the lesson and provide input from which others could learn. There were occasions when teachers would call on students who did not have their hands raised, or students who were not engaged appropriately, in order to 'keep them on their toes'. However, when the purpose of student discourse was to facilitate the learning of students, teachers called on those whom they believed would facilitate the learning goals of the class as a whole. Not surprisingly, those called on were regarded by the teacher as being the most able in science.

There is a possibility that males are more active when physical science-orientated topics are concerned and females are more active when topics which are life science-orientated are being studied. During the teaching of a topic on vertebrates, a number of grade 10 males remained in their seats and did not participate in an activity involving observation of animal skeletons. When members of the research team asked them about their non-participation, the males replied that they were not interested in the activity and could get answers to the questions from the book. In contrast, the same males were extremely active during a food testing activity that involved heating various foods and reagents. In fact, the males monopolised the equipment to such an extent that several females were unable to complete the activity at all. In the process of rushing through the activity, the reagents were contaminated and were unsuitable for use by others at a later time. Similarly, during a later topic on nuclear energy, the same males quickly commandeered a geiger counter and used it to test sources for possible radioactive emission. Thus, the tendency to get involved not only provided males with opportunities to learn because of their greater involvement in activities from which learning can occur, but it also diminished opportunities for female students to learn, not only because females were confined to a more passive type of involvement, but also because materials needed for laboratory activities were not available or were no longer appropriate.

On one occasion, a teacher set up a number of investigations on energy transformations around the classroom. There was sufficient equipment for everyone to be involved and sufficient time for them to use the equipment. Although there was evidence of males monopolising the use of equipment at various times during the lesson, almost everybody participated in each of the activities during the 100 minutes that were available. In this situation, gender differences were minimised by providing more than adequate time to complete the activities and by having enough equipment to allow everybody to have direct experience.

There were some striking counter-examples to the above results. In grade 11 in Australia, where science is an elective area of study, a small number of females was observed to dominate the interactions in some classes. Females who elected to study chemistry and physics were very capable and had selected these subjects because they were prerequisites for courses to be pursued at university. For example, Sue and Maria had decided to study physics because the medical school at a university recommended that physics be studied. During the physics lessons that were observed, each of these girls was active in whole-class interactions, seatwork and laboratory activities. Each was task-oriented, confident in her ability to succeed, and motivated to learn. During an interview, each indicated that she would respond to teacher questions and ask questions to facilitate understanding of physics. It is apparent, from their active involvement, that Sue and Maria could be regarded as voluntary minorities. They accepted science as it was and, despite the forces that hold out many females, they engaged in such a way that they could be successful and make progress towards their career goals.

The use of compliant females to dilute the effects of disruptive males was a gender-related phenomenon that operated in many classes to the disadvantage of females. This tendency brought with it a disadvantage for those students who were committed to learn and motivated to work without disrupting the learning of others. Instead of working with others who were committed and motivated to learn, they had in their group at least one person who was likely to be disruptive. In some cases, males who were disrupting a lesson were moved to a group that was functioning productively.

It is clear from our research that characteristics of students and the manner in which the teacher implements the science curriculum are associated with gender inequities. In the next section, I explain how a metaphor used to conceptualise the role of the teacher constrains how the teacher interacts with the students and leads to some of the females being disadvantaged. The possibility that gender differences can be overcome by changing the metaphors used to conceptualise teaching roles is explored.

GENDER INEQUITIES EMBEDDED WITHIN METAPHORS

Tobin *et al.* (1990) described how Peter, a high school science teacher, described his teaching role in terms of two metaphors, the captain of the ship and the entertainer. The classroom environment was a direct reflection of the metaphor Peter used to guide his teaching practices. For example, when Peter was captain of the ship, his practices as a captain constrained the possible roles for students in his class. In this role, Peter was a stern taskmaster who kept all students on task, regardless of gender. However, when Peter became an entertainer, he was more informal and interactive with students. In this role, gender differences became much more apparent. As an entertainer, Peter liked to interact with the more attractive females in his class. The nature of the interactions was different from those that occurred when Peter's teaching was guided by the captain of the ship metaphor. He was serious with some female students, he appeared to flirt with others, and he ignored some completely. Also as an entertainer Peter appeared to project himself as a male and deal with some of his students as females. Thus, in one style of teaching, gender differences were not apparent in the class and, in another predominant style, gender differences were pervasive. When Peter's teaching was framed by the entertainer metaphor, some females appeared to be advantaged and others were disadvantaged.

The females who appeared to be advantaged by Peter's entertainer style of teaching included a diligent worker who sat at the front and had numerous serious discussions with Peter about the science subject matter. The disadvantaged females were those whom Peter considered to be unattractive (e.g., one who was overweight) and those with whom he appeared to flirt. The former group did not have as many opportunities to interact with Peter about learning science, and females in the latter group tended to have interactions that were characterised by humour, sarcasm, sexual innuendo and light-hearted banter. Students in this category, often were caught in a transition when Peter switched from being an entertainer, with its associated rule structure, to being captain of the ship, with a more authoritarian, teacher-centred rule structure. In such a climate, it is not difficult to imagine that some female students would begin to see themselves as apart from a culture in which they were subjected to harassment by those in control. Thus, Peter's use of the entertainer metaphor inadvertently could have resulted in a classroom climate in which female students were constrained to begin or continue a journey towards becoming involuntary minorities.

If Peter acknowledges that gender equity is a problem which he needs to address, there is a possibility that he will adapt and create equitable learning opportunities for all. Alternatively, colleagues or researchers might discuss gender equity with Peter in terms of his entertainer metaphor and our classroom observations. This strategy appeals as having promise because Peter could develop tools to reconceptualise his roles in terms of new metaphors and, through reflective discussions, create narratives that, at a later time, could become foci for conversations about teaching, learning, gender equity, and the constraining effects of metaphors on the enactment of the curriculum.

Because so many of our metaphors develop from images associated with life outside of teaching, strategies can be framed in terms of a metaphor that, prior to its use as a referent for teaching, had little or nothing to do with teaching and learning science. These metaphors, such as the entertainer, could be based on images of popular television shows and inadvertently might carry embedded connotations of females as sex objects. An image that seems to fit, in the sense that a teacher feels right when he projects that image, might carry with it actions that demean female students and result in them being disadvantaged in the science classroom. Thus, becoming aware of the major referents that frame teaching is an important recommendation for all teachers. As a starting point, teachers might endeavour to identify the most salient images of self and non-self that characterise the classroom, identify metaphors and images used to frame roles of the teacher and students, and examine the involvement of all students with respect to the teachers' personal theories of learning. Our research suggests that several sets of beliefs are most salient in determining what happens in science classrooms. These include beliefs about learners, learning, self, teaching, control, knowledge and restraints. Discussion of what happens in the classroom in relation to beliefs about the factors mentioned above might lead to the creation of an environment that is conducive to the framing of new problems and the development of a commitment to personal change regarding factors that relate to gender equity.

CONCLUSIONS

The factors that determine whether or not males and females engage in the same way with science are complex, because of the joint influences of the teacher, the students and the culture in which the curriculum is embedded. Because of the inter-active effects of teacher and student variables, endeavours to change the patterns of student engagement probably should be directed to both teachers and students. A focus on the teacher, without including students in a fundamental way, might not succeed.

Teachers have power and exercise control over almost every aspect of classroom life. It is rare for students to have autonomy and opportunities to exercise control and choice in respect of matters related to their learning. Perhaps this is an aspect of life in classrooms that could change most readily. Is it possible for students to learn about their own learning, monitor their opportunities to learn, and discuss with peers how they might get more involved in activities likely to enhance their learning? Students who feel that they are disadvantaged might have opportunities to describe the nature of their disadvantage and mechanisms could be established to allow students to inform teachers when they are not being provided with the same opportunities to learn as others in the class. Just as much of what happens in a classroom is invisible and natural for teachers, so too will it be for students. Thus, if students are to be reflective on their opportunities to learn science, it is important that they construct viable theories of learning science. Empowering students to monitor and discuss their own opportunities to learn holds potential for the reform of what happens in classrooms. By alerting students to the possibility of gender inequities and em-powering them to identify problems when they are perceived, it is possible that the teacher and all students will become sensitised to creating equal opportunities to participate and learn and will take corrective actions when inequities are substantiated.

To the extent that schooling facilitates the journey of many females towards be-coming involuntary minorities in science and then refugees away from science, it is essential that efforts are made to provide all students, including females, with high-quality learning experiences. At the present time, scarce resources and a curriculum that still reflects some of the beliefs and practices that were more appropriate for the beginning of the 20th century combine to make science, as it is enacted in schools, inaccessible to most students, including a disproportionate number of females. The incorporation of learning resources from the home and community at large might provide avenues for making science learning activities more inclusive. Teachers might encourage female students to use modems to hook into the internet and access learn-ing resources that will facilitate learning and build interests in an environment where competition with male peers is not an issue. However, if students continue to be organised for science learning in classrooms with one teacher and 30–40 peers, then efforts will need to be directed towards building environments that are conducive to the learning of all students. Teachers and students should look carefully for the signs of involuntary minorities, and when they discern them, bring them to the attention of all. Such signs should not be permitted to flourish as they have done in the past. In an era of standards-based reform in which science for all is advocated, we no longer should tolerate inequities associated with gender in our classrooms. All teachers and students should embrace equality as a reality of the 21st century, if we cannot achieve gender equity before then.

Florida State University, USA

REFERENCES

Kahle, J.B. & Meece, J. (1994). 'Research on gender issues in the classroom', in D.L. Gabel (ed.), *Handbook of research on science teaching and learning*, New York, Macmillan, 542–557.
Ogbu, J.U. (1992). 'Understanding cultural diversity and learning', *Educational Researcher* 21(8), 5–14 & 24.
Tobin, K. & Espinet, M. (1989). 'Impediments to change: An application of peer coaching in high school science', *Journal of Research in Science Teaching* (26), 105–120.
Tobin, K. & Gallagher, J.J. (1987). 'What happens in high school science classrooms?', *Journal of Curriculum Studies* (19), 549–560.
Tobin, K., Kahle, J. & Fraser, B. (eds.) (1990). *Windows into science classrooms: Problems associated with higher-level cognitive learning*, London, Falmer Press.

11. EQUITABLE SCIENCE EDUCATION: A DISCREPANCY MODEL

A 15-year-old girl in rural America described the current crossroads in science when she said: 'There are some women scientists; but men have been in it longer. Women can do the same job as men. They may have a different way of thinking and might improve science' (Kahle, 1985, p. 68). Her words were fortuitous because they were spoken a few days before Barbara McClintock won the Nobel Prize for looking at maize in a different way and for thinking about genetics in a different manner. McClintock's work, unrecognised and even scorned for decades, epitomises the disadvantages that not only individual women but also the scientific community and society as a whole suffer because of a lack of equity in science education. Perhaps Maria Mitchell, one of the first American women to be recognised as a scientist, said it best:

In my younger days, when I was pained by the half-educated, loose, and inaccurate ways which we (women) all had, I used to say 'how much women need exact science', but, since I have known some workers in science who were not always true to the teachings of nature, who have loved self more than science, I have now said 'how much science needs women'. (Maria Mitchell's presidential address to the Third Congress of Women in 1875; quoted in Rossiter, 1982, p. 15)

Over a century later, as the McClintock story dramatised, science still needs women. Yet, study after study in the developed, Western world suggests that girls and women receive very different educations in science than do boys and men.

THE DISCREPANCY MODEL

Progress towards equitable science education can be analysed by using a discrepancy model to distinguish between ideal and actual states. First, the ideal state is constructed from objective evidence; next, the actual state of science education is described. And, last, the transition from actual to ideal state is delineated. In developing the model, a search of the work of experts from a variety of fields (e.g., science, education, sociology and psychology) is a prerequisite. The ideal state cannot be based on opinion; rather, it must be developed from the vast array of research available. The actual state, too, is built from the writings of a variety of experts, including classroom teachers and educational researchers. Then, one examines the differences between the two states with the goal of determining ways to progress from the actual to the ideal state or from inequitable to equitable education in the sciences.

Before constructing our discrepancy model, it is important to define equitable education. Generally, education is equitable when all children participate and achieve equally. Some have referred to it as 'sex fair education', while others have maintained

L.H. Parker et al. (eds.) Gender, Science and Mathematics, 129–139.
© 1996 *Kluwer Academic Publishers. Printed in the Netherlands.*

that it is achieved by a 'gender-inclusive curriculum'. For example, the Australian Science Teachers' Association discusses equitable education in terms of the curriculum, and it defines 'gender-inclusive curriculum' as curriculum in which content, language and methods give 'as much value and validity to the knowledge and experiences of girls and women as that given to boys and men' (Rennie & Mottier, 1989, p. 18). Calabrese (1988) describes 'sex fair education' as education which recognises that all students have different personal needs regardless of their sex. Therefore, equitable education in science is based upon a curriculum that includes the experiences and needs of all students and encourages equal participation and outcomes for girls and boys.

THE IDEAL STATE OF SCIENCE EDUCATION

Reaching the ideal state is hampered by the popular image of science (described in the actual state). In the ideal state, science will have a factual rather than a romanticised image; it will be characterised not as the objective discovery of truth but rather as a very human and humane endeavour. Understanding will be emphasised rather than absolutes. Creativity and social discourse will become an integral part of school science as they are in actual science. Cooperative group learning in schools can, and should, illustrate the ways in which laboratory groups of scientists identify, probe and solve problems.

Primary School

Many have argued that the primary school is the critical place for change: change in formal and informal science curricula, change in classroom instruction and interactions, and change in school structure and socialisation.

The Curriculum. An anomaly exists in that examples of ideal primary science curricula have existed for several decades, yet they are little used. In the USA, examples include: SCIS, Science Curriculum Improvement Study; SAPA, Science, a Process Approach; and ESS, Elementary Science Study. Those curricula were developed in the 1970s and have faded from classrooms for three primary reasons: teachers have not understood the scientific principles which the materials promulgated; classrooms have not been organised for small-group interactions in science; and schools have not provided the equipment or the scheduling required. However, recent analyses make a strong case for their redemption in the ideal state. For example, a review of 34 evaluative studies indicates that children using the activity-based curricula achieved better on every measure of achievement (science processes, creativity, perception, logic development, science content and mathematics) than children studying 'textbook' science (Shymansky *et al.*, 1983). Other research has shown that opportunities for experimentation, for handling instruments, for making measurements, for observing natural phenomena, for collecting data and for making interpretations have the potential to produce equitable education for all children (Kahle & Rennie, 1993). Because girls and boys enter primary school with equal interest, but unequal experiences, in science (Kahle & Lakes, 1983; Kelly, 1985) the activity-based curricula provide girls with experiences that they are less likely than boys to bring to school.

The Classroom. Partly because of the use of process curriculum and partly because of proposed changes in the education of primary teachers, instructional interactions and teacher behaviours will be different in the ideal state. The suggested changes are grounded firmly in the findings of recent research. For example, in comparison with girls, boys more often use science instruments and materials (Kahle & Lakes, 1983); boys read more science-related books (Kahle & Lakes, 1983); boys interact with teachers more frequently (Smail, 1984; Tobin *et al.*, 1990); boys are assigned higher grades on science papers (Spear, 1984); and boys are asked more higher-order cognitive questions in science lessons (Tobin *et al.*, 1990). All of those findings indicate changes that are needed to make primary science lessons more equitable.

Some have argued that teachers will be unable to make the prerequisite changes because girls themselves will resist them. However, one study analysed the number of times that girls responded that they had used the equipment, tools and instruments of science (actual activities) with the number of times that they wished to use them (desired activities). Girls wished to do significantly more and varied science activities than they actually had done (Kahle & Lakes, 1983). Teachers, cognisant of such studies, will insist that girls actively participate in science activities. That is, in the ideal state, teachers will expect comparable work and behaviour from both boys and girls and will reward both in similar ways.

The School. Changes also are needed in the primary school as an institution. Basically, convenient management techniques which consistently separate children on the basis of sex will be avoided. On both the playground and in the classrooms, areas will be specified for quiet and adventurous activities, and both boys and girls will be encouraged to participate in both types. Older boys as well as girls will be assigned to assist with younger children. The school in the ideal state will provide an optimal setting for science instruction because there will be equity in the corridors as well as in the classrooms.

Secondary School

In the ideal state, secondary schools and their classrooms of biology, geology, chemistry and physics will be pivotal in terms of retaining students in science. The following description of ideal secondary science curricula, classes and schools is derived from the findings of recent studies in the USA, the UK and Australia.

The Curriculum. Recently, both researchers and practitioners have identified aspects of the curriculum that broaden its appeal and its potential for effectiveness. For example, in the UK, Garratt (1986) has suggested four factors which influence the subject choice of both girls and boys: interest (significantly more important for girls than for boys); previous performance; career value; and, to a lesser extent, relevance to everyday life.

Eccles (1989) uses several mediating factors in developing a model to explain how students make subject choices. Those mediating factors include causal attribution patterns for success and failure, the input of socialisers (primarily parents and teachers), gender-role stereotypes and one's perceptions of various possible tasks. Each of the mediating factors contributes both to the perceived value of the activity and to one's expectation for future success in it. According to her model, expectations for success and perceived value influence the amount of effort that one will expend on an activity, the performance level that will be achieved, and the decision to participate or to continue in the activity.

To test her model, Eccles and her colleagues assessed the effect of expectations and values on the students' decisions to pursue mathematics and English courses in the USA. They found that differences in subject choice patterns became apparent in junior high (ages 12 to 14 years) and increased throughout high school. During those years, girls became relatively more confident about their English skills and less confident about their skills in mathematics, even if they were doing well in mathematics courses. Furthermore, their perceived value of mathematics decreased also. Similar changes were not apparent for secondary boys. Although gender-role expectations affected both boys' and girls' expectations of success and perceived value of subjects, girls and boys reacted to those expectations and values in different ways. For example, high school girls were more likely than boys to choose advanced courses in subjects about which they were more confident and which they valued (English, foreign languages and social sciences, rather than mathematics or physical science courses). In the ideal state, secondary science curricula must include the interests of girls, reflect the value of studying science for girls, and enhance girls' as well as boys' expectations for success in science.

Two intervention projects, one in the UK and the other in the USA, focussed on curriculum that appealed to both boys and girls (gender-inclusive), and their findings provide a partial description of science curriculum in the ideal state. The USA study identified and observed biology teachers who were successful in motivating tenth grade girls to elect optional physics and chemistry courses (Kahle, 1985). Its findings provided a composite picture, as well as a collective pool of data from which commonalities were identified and generalisations were made. In the UK, the Girls in Science and Technology (GIST) project involved 10 comprehensive schools in the Manchester environs and 2,000 children from the time when they entered lower secondary school (age 11 years) until when they made their subject choices at age 14 years (Smail, 1984). Although its findings portrayed the actual classroom and school, the research team developed some ideal curricula and hypothesised about ideal situations.

GIST proposed that an ideal lower secondary school science curriculum would be constructed around girls' and boys' expressed interest in the human body. It also recommended that the contributions of women scientists be integrated into the curriculum. In addition, the inclusion of 'tinkering' activities in order to overcome the lack of such experiences by girls was stressed. The results of both the GIST and the USA project demonstrated that an ideal secondary science curriculum must provide experiences with rotating three-dimensional figures in space, with drawing and conceptualising three-dimensional forms, and with projecting curvilinear distances and outcomes. These experiences enhance a student's visual-spatial acuity. Because girls usually have fewer experiences with the toys, games and activities which develop that attribute, opportunities must be provided within the curriculum. The GIST project revealed that, although boys initially scored better on spatial ability tests, the enrolment of girls in one technical craft course eradicated the gender difference (Smail, 1984).

Extensive laboratory work will be needed in the ideal curriculum, because laboratories enhance both interest and experience. The need for experience with the actual tools and techniques of science is supported by the findings of both studies. Boys as well as girls express anxiety if they do not have sufficient past performance against which to gauge future success. For example, the USA study found that girls express little anxiety about focussing a microscope with which many have had experience, but great anxiety about wiring an electric circuit for which they have had no experience. Boys, on the other hand, express concern about taking the temperature of a living organism, a technique with which they have had less previous experience.

The USA study suggested that science curriculum in the ideal state will include alternative and supplementary materials as well as teacher-developed materials which include examples drawn from the common experiences of girls (sewing machines and volley ball) as well as those of boys (cars and football). In addition, the USA study indicated the value of explicit career and educational counselling by science teachers.

The Classroom. Many studies have analysed teacher behaviours and instructional strategies, and their findings allow us to describe science classes in the ideal state. Ethnographic studies especially have yielded rich data. Classroom observations reveal that, although teachers give boys specific instructions for completing a problem, they often show girls how to finish a scientific or technical task. Furthermore, when interacting with boys, teachers are more likely to use the strategies of leaning forward, looking into eyes, nodding and smiling to indicate anticipation of a superior performance. Both verbal and nonverbal interactions can communicate to boys and girls differential teacher expectations that are group based (i.e., achievement expectations which are a function of one's sex and/or race). Because students cannot change their sex and/or race, they accept the achievement expectation as something which they cannot change.

The USA study discussed above focussed primarily on instructional strategies which encourage girls as well as boys to continue to study science; its findings provide some guidelines for science instruction in the ideal state. First, although it will emphasise classroom discussion as well as individual and laboratory work, cooperative group work will be the primary learning strategy. Second, diverse media as well as field experiences will be used. Third, independent projects and library research will be integrated into the instruction. Furthermore, three unique findings of the USA study portend changes in instructional patterns. First, teachers, who were successful in encouraging students to continue in science quizzed or tested their students once a week. Second, they encouraged creativity. Third, they fostered basic skill development (Kahle, 1985). The implementation of science instruction which develops creativity and originality while teaching basic skills will be an important change in the ideal science classroom. The infusion of creativity will add a new dimension to science instruction, while competency in basic skills such as measuring, graphing and titrating, will increase the probability of success in science for all students.

The School. Secondary schools in the ideal state will identify and encourage creative science teachers. In addition, they will provide both a setting and administrative services to foster learning free of sex-role stereotypes. Courses will be scheduled so that students can take as many science options as they wish. In addition, traditionally female and male classes will not conflict so that girls, too, will enrol in electronics, drafting, metal working, physics, etc. Furthermore, counselling or guidance offices within schools will provide non-traditional information rather than promulgate sex-stereotyped course and career selections. Changes within the counselling system will be among the most important ones, and informed choice will be the key for both boys and girls in the ideal school.

THE ACTUAL STATE OF SCIENCE EDUCATION

Other contributions to this volume have demonstrated that today, in the minds of our students and in the perceptions of their teachers, science is masculine. Indeed, the scientist has replaced the cowboy in the adolescent's imagination as the hero, or anti-hero, who is fearless, strong and lone. It is futile to argue that, because that image could be derived largely from the media (i.e., television, movies and advertisements), we are helpless to change it. There is a wealth of evidence to support the contention that, regardless of how, when and where the masculine image of science evolved, schools and universities as well as teachers and professors sustain it. In fact, gender often is recontextualised within schools so that academic disciplines become stereotyped. Once a subject has acquired a gender status, in this case masculine, participation in it is seen to reinforce a boy's masculinity and to diminish a girl's femininity. Indeed, Kelly (1985) maintains that '[T]he masculinity of science is often . . . the prime reason that girls tend to avoid the subject at school'. She suggests that 'schools could play a transformative, rather than a reproductive role, in the formation of gender identities' (p. 133).

As indicated in Chapter 1, Kelly's (1985) four factors that contribute to science's masculine image include: the numbers of men and women who do science and who receive recognition for their work (Nobel laureates, fellowships, research grants); the packaging of science curriculum (examples, exemplars); and the practice of science teaching (sex-role stereotypes, student-teacher interactions). Those factors will be used to describe the actual state of science education.

Numbers

In primary education, the real issue is not the overt number of students studying science, but the covert ways in which boys and girls experience science. Today, boys and girls bring different science experiences to school, and they receive very different science education in school. For example, there is clear documentation that fewer girls than boys handle science equipment, perform science experiments or participate in science-related activities (Kahle & Lakes, 1983; Parker & Rennie, 1986; Smail, 1984).

The image of science and science courses becomes more masculine in secondary school where the numbers of boys taking science and of men teaching science increase. In the USA and Australia, although women are well-represented among secondary biology teachers, teachers of the physical sciences are predominantly male. Further, the ratio of male to female teachers is reflected in the ratio of students enrolled in various courses. In England, Kelly (1985) reported that boys constituted 70–80% of all examination entries in physics, 60% of all candidates in chemistry and only 30–40% of the biology examinees. In the USA, the pattern is similar. Although virtually all high school students take biology, which functions as a required, introductory science course, approximately 45% of university-bound high school girls take chemistry, while physics, taken by 51% of university-bound high school boys, is studied by only 35% of university-bound American girls (Czujko & Bernstein, 1989). Likewise, Australian enrolment data for grade 12 science courses shows that, although the proportion of girls studying chemistry increased from 29% in 1976 to 40% in 1987, the proportion of girls studying physics rose only 5% in that same period (from 22% to 27%). Furthermore, although over 25% of Australian boys meet tertiary entrance requirements with two mathematics, physics and chemistry units, only 6% of Australian girls have comparable records (de Laeter *et al.*, 1989).

Packaging

Texts, published materials, posters, library books, and exemplars all portray more male scientists and incorporate more of their work. In the USA, publisher guidelines ensure that segments of its population are represented pictorially in correct proportions. Therefore, 50% of illustrations and diagrams show females and 17% depict African-Americans. However, the cosmetic changes mask the lack of substantive ones (i.e., substantive discussion of women's scientific contributions). Rennie and Mottier (1989), whose analysis of Australian science texts presents similar findings, recommend that both text and illustrations show men and women participating equally in science.

Practice

The actual practice of science at all educational levels contributes to its masculine image. For example, when primary teachers are asked to identify scientifically talented or gifted students, a pattern emerges. Both Australian and American teachers identify more boys. When observers record both the number and duration of teachers' interactions with the identified creative girls and boys, they find that teachers interact twice as often with the boys and for longer duration. In England, Spear (1984) has analysed the marking of science papers attributed to 12-year-old boys and girls and has found that more male and female science teachers give high marks when the work is attributed to a boy. In addition, in today's science classes, boys often are allowed to dominate discussions and equipment and are four times more likely to be target students (i.e., students who dominate verbal interactions with the teacher) than are girls (Tobin *et al.*, 1990). The actual practice of science today as factual knowledge, presented by whole-class instruction and related to male activities, is an anathema to girls both at the primary and secondary levels.

TRANSITION FROM ACTUAL TO IDEAL STATE

The changes needed to move from the actual to the ideal state are based on the premise that changes in the practices and packaging of science will affect the numbers of women who do science and, therefore, diminish its masculine image. As science's image becomes more accurate, the numbers of girls participating and achieving in science will increase, and science classes will transform, not reproduce, society's stereotypic vision of science. As Kelly has said:

[T]he masculinity of science is an *image*. Whether it is caused by textbook representations or by classroom behaviors, it is essentially a distortion of science. The word 'image' is linked to 'imaginary' and those three mechanisms all suggest that the masculinity of science is only an illusion, not an intrinsic part of its nature. (Kelly, 1985, p. 146)

During the last two decades, many projects have focussed on enhancing girls' interest in science, informing girls about careers in science, and improving their attitudes about science. Unfortunately, those projects produced only limited results. For example, Vetter (1987) reports that the number of USA women enrolling in university science, mathematics and engineering majors peaked in 1984. Data from the National Science Foundation (NSF, 1992) indicate that, since the mid-1980s, freshman women have been less inclined to pursue scientific or technological majors. Interest-based projects may have increased girls' motivation to study science without providing them with the skills to do science. Although girls enter the pipeline to scientific careers, they subsequently leave — usually during physics or calculus courses at the secondary or tertiary levels.

If we compare the progress that has been made in encouraging girls in science with what has been accomplished in mathematics, some needed transitional steps become clear. Projects designed to increase the numbers of girls and women studying mathematics have included career information and have addressed mathematics anxiety. Furthermore, their primary focus has been on student skill levels in mathematics. Therefore, problem-solving strategies, as well as skills in using logic and approximation, have been stressed. Mathematicians and mathematics educators in both the USA and Australia have incorporated into the curriculum activities that develop students' visual-spatial abilities (AAMT, 1990; Ben-Chiam *et al.*, 1988). In other words, projects designed to encourage girls in mathematics have moved beyond information about careers and role models to activities that enhance and normalise students' skills in doing mathematics. As a consequence, in the USA, the number of mathematics courses taken by high school girls continues to rise; in 1988, girls completed 3.6 years, compared to 3.8 years for boys. Furthermore, the percentage of girls completing calculus in USA secondary schools has risen from approximately 10% to 15% in the last decade.

Recent intervention projects and research in science education have reinforced the findings from mathematics. In mathematics, Ben-Chiam *et al.* (1988) found that short-term instruction with visual-spatial activities eradicated differences in favour of boys on a visual-spatial ability test. In the USA, a team of researchers and biology teachers have worked together to improve the achievement levels and retention rates of students in science by focussing on enhancing their skills in doing science (Kahle & Danzl-Tauer, 1991). A unit in genetics was selected by the teachers, and daily exercises which focussed on visual-spatial skills (building and using three dimensional models as well as designing and building laboratory apparatus) and on quantitative skills (collecting and transforming data) were developed by the research team. Teachers were an integral part of the study team and were free to select and use the developed activities as well as a variety of other resources that focussed on genetics and on women's contributions to and participation in science. Results indicate that both achievement levels and future subject choices were affected positively for both girls and boys in classes that incorporated consistently the skill activities into the curriculum.

During the transition from actual to ideal state, science teachers and professors must be agents of change. Again, two projects suggest why. Smail (1985) has suggested that the results of the GIST project were limited because many of the teacher-participants did not 'buy into' the program. Likewise, a USA project that provided a smorgasbord of equity and scientific resources and materials to teachers produced no significant changes in achievement and attitudes towards science (Kahle, 1987). Although the teachers were involved actively in both projects, they were neither an integral part of the research team nor highly involved in the decision-making processes concerning project activities.

Because the transition phase will rely on teachers, changes in both their practice and preparation will be needed. One project in the USA developed a semester-long course on equity issues in science for primary teacher trainees (McDevitt *et al.*, 1993). Although initially students entering the course (almost all women) questioned its need and premise, course evaluations indicated that changes occurred in their attitudes towards science, in their confidence in teaching science and in their proposed techniques for teaching science. The researchers, however, found that positive support of a cooperating teacher was needed for the students to transfer their skills and concerns into action during their field experiences. Therefore, they recommended concomitant training in equity issues and skills of science for practising teachers.

A second USA study analysed the successful incorporation of equitable teaching strategies by teacher trainees during their teaching practicum (Scantlebury & Kahle, 1993). The results indicated that secondary preservice science teachers placed with cooperating teachers, who also were trained in equitable teaching strategies, were successful in implementing those strategies. In comparison, trainees who participated in a peer support group during practice teaching did not use as many of the identified strategies. The transition phase will require both changes in and connections between preservice and inservice science teacher education.

SUMMARY

The ideal state portrays science education developed and practised within a realistic image of science. Simply changing science's image will do much to advance equitable science education; as the packaging and practices change from kindergarten through college, the numbers of girls and women in science will increase and the image and practice of science will change. Although there is neither a quick nor an easy route from the actual to the ideal state, a focus on all levels of education will speed the process. Several paths have been proposed for the transition; they are based upon research and projects that have been successful in promoting equitable science education. Science education, both in training and in practice, can benefit from the changes that mathematics education has undergone in the USA. Clearly, in science, as well as in mathematics, equitable education benefits all children, enables previously under-represented groups to perform well, and diminishes sex-role stereotyping of subjects and careers.

Miami University, Oxford, Ohio, USA

REFERENCES

Australian Association of Mathematics Teachers, Inc. (AAMT) (1990). *A national statement on girls and mathematics*, Adelaide, Australian Association of Mathematics Teachers.
Ben-Chiam, D., Lappan, G. & Hourang, R.T. (1988). 'The effect of instruction on spatial visualization skills of middle school girls and boys', *American Educational Research Journal* (25), 51–71.
Calabrese, M.E. (1988). 'What is sex fair education?' in A.O. Carelli (ed.), *Sex equity in education*, Springfield, IL, Charles C. Thomas, 75–81.
Czujko, R. & Bernstein, D. (1989). *Who takes science?*, A report on student coursework in high school science and mathematics, American Institute of Physics, New York.
de Laeter, J., Malone, J.A. & Dekkers, J. (1989). 'Female science enrolments in Australian senior secondary schools', *Australian Science Teachers' Journal* 35(3), 23–33.

Eccles, J.S. (1989). 'Bringing young women to math and science', in M. Crawford and M. Gentry (eds.), *Gender and thought: Psychological perspectives*, New York, Springer-Verlag, 37–57.

Garratt, L.L. (1986). 'Gender differences in relation to science choice at A-level', *Educational Review* 38(1), 67–76.

Kahle, J.B. (1985). 'Retention of girls in science: Case studies of secondary teachers', in J.B. Kahle (ed.), *Women in science: A report from the field*, London, Falmer Press, 49–76.

Kahle, J.B. (1987). 'SCORES: A project for change?', *International Journal of Science Education* (9), 325–333.

Kahle, J.B. & Danzl-Tauer, L. (1991). 'The underutilized majority: The participation of women in science', in S.K. Majumdar, L.M. Rosenfeld, P.A. Rubba, E.W. Miller and R.F. Schmalz (eds.), *Science education in the United States: Issues, crisis, and priorities*, Philadelphia, PA, Pennsylvania Academy of Science Press, 483–503.

Kahle, J.B. & Lakes, M.K. (1983). 'The myth of equality in science classrooms', *Journal of Research in Science Teaching* (20), 131–140.

Kahle, J.B. & Rennie, L.J. (1993). 'Ameliorating gender differences in attitudes about science: A cross-national study', *Journal of Science Education & Technology* 2(3), 321–334.

Kelly, A. (1985). 'The construction of masculine science', *British Journal of Sociology of Education* 6(2), 133–153.

McDevitt, T.M., Heikkinen, H.W. Alcorn, J.K. Ambriosio, A.L. & Gardner, A.L. (1993). 'Evaluation of the preparation of teachers in science and mathematics: Assessment of preservice teachers attitudes and beliefs', *Science Education* (77), 593–610.

National Science Foundation (NSF) (1992). *Women and minorities in science and engineering*, NSF no. 92-303, Washington, DC, National Science Foundation.

Parker, L.H. & Rennie, L.J. (1986). 'Sex-stereotyped attitudes about science: Can they be changed?', *European Journal of Science Education* (8), 173–183.

Rennie, L.J. & Mottier, I. (1989). 'Gender-inclusive resources in science and technology', *Australian Science Teachers' Journal* 35(3), 17–22.

Rossiter, M.W. (1982). *Women scientists in America: Struggles and strategies to 1940*, Baltimore, MD, Johns Hopkins University Press.

Scantlebury, K. & Kahle, J.B. (1993). 'Implementation of equitable teaching strategies by preservice teachers', *Journal of Research in Science Teaching* (30), 537–547.

Smail, B. (1984). *Girl-friendly science: Avoiding sex bias in the curriculum*, London, Longman.

Smail, B. (1985). 'An attempt to move mountains: The "Girls in Science and Technology" (GIST) Project', *Journal of Curriculum Studies* (17), 351–354.

Shymansky, J., Kyle, W. & Alport, J. (1983). 'The effect of new science curriculum on student performance', *Journal of Research in Science Teaching* (20), 387–404.

Spear, M.G. (1984). 'Sex bias in science teachers' ratings of work and pupil characteristics', *European Journal of Science Education* (6), 369–377.

Tobin, K., Kahle, J.B. & Fraser, B.J. (eds.) (1990). *Windows into science classrooms: Problems associated with higher level cognitive learning in science*, London. Falmer Press.

Vetter, B.M. (1987). 'Women's progress', *MOSAIC* 18(1), 2–9.

SECTION III

FROM POLICY TO PRACTICE –
BUILDING ON EXPERIENCE

THOMAS R. KOBALLA, JR.

12. THE ROLE OF PERSUASIVE COMMUNICATORS IN IMPLEMENTING GENDER-EQUITY INITIATIVES

At a time when gender-equity is a celebrated issue, it is unsettling to acknowledge that gender differences in science and mathematics interest and achievement continue. The small number of women who study science and mathematics in high school and beyond, the low representation of women in scientific and technological careers, and some data showing lower levels of female interest and achievement in science and mathematics have generated much debate among the experts who try to explain these events. Physiological explanations for the conspicuous gender differences in science and mathematics have been proffered (Rudisill & Morrison, 1989). However, most researchers tend to agree that the observed gender differences are due to culturally-imposed stereotypes (Hensel, 1989; Kahle & Meece 1994).

There is little that can be done today to alter any physiological factors that can affect gender differences in science and mathematics but, by providing an optimally stimulating environment during the early formative years, Hensel (1989) believes that a parent or teacher can assist a child in realising his or her scholarly potential. Unfortunately, through their intentional and unintentional differential treatment of girls and boys and through their gender-related expectations, parents and teachers tend to foster stereotypic attitudes and gender role standards in children and adolescents (Kahle & Meece, 1994). To counter these lamentable patterns, individuals and organisations desiring change have developed initiatives to raise the consciousness of all to gender-equity issues (see Kahle & Meece, 1994).

The focus of these initiatives is not purely informational. At their heart is the desire to persuade the participants to modify their own biases and differential expectations for girls and boys. To achieve these results involves more than the logical organisation of the information to be communicated. The characteristics of the source of a persuasive message must be considered. This chapter addresses the immediate applicability of research on persuasive source characteristics to implementing gender-equity initiatives. Primary attention is given to describing the nature of the persuasive source and situations in which source characteristics can be key to persuasion. Ways to increase the likelihood of persons interested in implementing gender-equity initiatives to be persuasive also are discussed. Here, persuasion is defined as the 'conscious attempt to bring about a jointly developed product common to both source and receiver through the use of symbolic cues' (Koballa, 1992, p. 67).

L.H. Parker et al. (eds.) Gender, Science and Mathematics, 143–154.
© 1996 *Kluwer Academic Publishers. Printed in the Netherlands.*

NATURE OF PERSUASION

Logic suggests that, if a highly credible, attractive or powerful source were used to persuade someone, the level of agreement between the source and receiver would increase. But research has shown that this does not happen always. The person was persuaded in some studies (Chaiken, 1979) but not in others (Rhine & Severance, 1970), and in some cases the person's attitude was less positive after the message was delivered (Hass & Grady, 1975). Even when persuasion was successful, sometimes it would endure and other times it would not (Petty & Cacioppo, 1986).

Such disappointing findings led researchers to abandon the premise that had guided the research since the 1940s – that message learning results in persuasion – and to the realisation that it is not remembering the message, but the active construction of personal thoughts related to the message that leads to persuasion. This cognitive response view of persuasion fostered research that provided evidence for the notion that effects linked to source characteristics also can be mediated by message-relevant thinking. Studies couched in this view of persuasion led Richard Petty and John Cacioppo (1986, 1990) to postulate that some people actively were processing information presented in messages while others were not.

Petty and Cacioppo's work in this area resulted in the development of the Elaboration Likelihood Model of Persuasion (ELM) that proposes a continuum that reflects the extent to which a person analyses the arguments contained in a persuasive message. The continuum is anchored at one end by a central route to persuasion and at the other end by a peripheral route to persuasion. When persuasion is induced through careful and thoughtful consideration of the true merits of an issue, according to Petty and Cacioppo, central processing is occurring. On the other hand, when people do not have the motivation or the ability to process the message, their focus shifts to peripheral cues – like source characteristics or the length of the message – that become important determinants of persuasion. Interestingly, children were found to be persuaded more easily by appeals based on peripheral cues than on issue-relevant arguments (Miller et al., 1975), presumably because they have yet to acquire the cognitive skills to analyse issue-relevant information critically.

Tests of the ELM showed that argument quality was the most important persuader for people who viewed the topic of the message as personally relevant and that source characteristics influenced their persuasion more often when the topic was not viewed as relevant (Petty et al., 1983). For topics considered neither high nor low in personal relevance, only when strong arguments were presented did messages attributed to expert and attractive sources enhance persuasion (Heesacker et al., 1983; Puckett et al., 1983).

Based on the work of Petty, Cacioppo and their colleagues, it seems that the operation of source characteristics is not as simple as it might appear at first. In dealing with the real world issue of gender equity in science and mathematics, it is most likely that the personal relevance attached to the issue differs from one person to the next. Some people might consider the issue of gender equity in science and mathematics to be highly relevant, while some might consider it not relevant at all, and still others might view the issues as moderate in personal relevance. Especially for the latter two conditions, the characteristics attributed to message sources regarding gender-equity initiatives in science and mathematics take on added importance when considering persuasion.

MESSAGE SOURCE

One's ability to influence others to implement gender-equity initiatives could come from several sources. These bases of persuasion depend on whether or not the message recipient perceives the source as possessing qualities that are considered relevant to the goal of gender equity in science and mathematics. Three sets of characteristics hypothesised to affect the persuasive impact of a message regarding gender equity, when they are attributed to its source, are credibility, attractiveness and power. While each of these qualities can increase the amount of attitude change produced by a source, each brings about persuasion in a slightly different way.

Kelman (1961) first conceptualised these categories of source characteristics, the psychological mechanisms through which they operate, and their attitudinal consequences. While conceptually distinct, the three categories can be mixed in reality into a single source. An illustration, inspired by R. Glen Hass (1981), of the difficulty in separating the three categories of source characteristics in real world experiences might be helpful. Ms Jones, the high school science department head, often talks with Mr Woods, a teacher in her department, in the laboratory preparation room as they prepare for the next day's classes. Several times during the term, Ms Jones stresses the importance of employing practices that foster gender-equity in science classes to Mr Woods, who seems to be lukewarm about the idea.

One day, Mr Woods enters the preparation room and reports to Ms Jones that Mr Smith, a fellow science teacher, does not promote gender equity in his classes because he only calls on boys to answer higher-order questions and rarely chooses girls to help with experiments. Ms Jones quickly agrees that Mr Smith should also ask girls higher-order questions and choose them to help with experiments, while continuing to prepare for the next day's classes. Ms Jones is pleased that the attitude which Mr Woods has expressed agrees with the position that she has advocated in previous talks.

As one thinks about the persuasive success of Ms Jones, one can only speculate about the source of her influence. Does it result from one or a combination of all three characteristics – her credibility, attractiveness or power as department head?

Source Credibility

Credibility refers to the overall believability of the source and is considered to involve two basic factors – expertise and trustworthiness (Hovland *et al.*, 1953). A source projects expertness to the extent that she is perceived to possess valid and correct information. Trustworthiness is linked to the perception of the source's intent to communicate that information without prejudice. Kerr *et al.* (1982) describe communications from an expert and trustworthy source in terms of the concepts of validity and reliability. Communications are valid, they say, if they present relevant data, and they are reliable to the extent that they are consistent and objective. Thus, a credible source is one who is perceived to know what she is talking about and who has the best interests of the audience at heart.

Communicator credibility related to specific science and mathematics education issues has received some attention. Several reports suggest that scientists (Martin, 1985), science teachers (Koballa, 1988), university methods instructors (Shrigley, 1976) and curriculum supervisors (Shrigley, 1980) function as expert communicators on gender-related issues. For early adolescents, parents also can serve as trustworthy sources of information regarding gender-related issues (Koballa, 1988). Parental encouragement and support often are associated with adolescent's decisions to enrol in elective physical science courses in high school (Koballa, 1988). Teachers, counsellors and peers also can operate as trusted advisers in decisions about science and mathematics courses and careers (Kahle & Meece, 1994; Trigg & Perlman 1976).

'Internalisation' is the name given by Kelman (1961) to the psychological process by which credibility produces attitude change. Internalisation occurs, according to Hass (1981), when people accept a new attitude as their own by integrating it into their belief and value system. Internalisation is the process most often thought of when attitude change is mentioned and is associated with enduring change, even after the identity of the message source is forgotten.

Reflecting on the talks between Ms Jones and Mr Woods, one might wonder if the gender-equity attitude expressed by Mr Woods is one which he had internalised. If internalised, the attitude is likely to be due to Mr Woods' awareness of Ms Jones' knowledge of barriers to gender equity, her experience in directing gender-equity initiatives, or perhaps her motivation to promote equal opportunity for males and females in science and mathematics courses and careers.

Source Attractiveness

Attractiveness refers to the personal qualities that make the message source a person with whom to identify. To the extent that a source is liked or admired, audience members can choose to adopt the same attitude as the source. Kelman (1961) used the term 'identification' to characterise this psychological process, and he described it in terms of a yearning to create gratifying role relationships with the message source by the message receiver. Although little is understood clearly about how attraction operates in long-term relationships, much is known about the factors that lead to initial attraction.

Various factors seem to account for why a source is considered attractive by an audience. For example, good looks (Chaiken, 1979), familiarity (Saergert *et al.*,

1973), or similarity in appearance or attitude (Byrne, 1971) increase a person's attractiveness and persuasiveness. In a study of source attractiveness, Chaiken (1979) found that people considered physically attractive and unattractive also differed in ways other than appearance. Attractive sources had more positive self-concepts and better communication skills than their unattractive counterparts. These additional factors are thought to have contributed to the different persuasive effects observed between the attractive and unattractive sources. Studies of communicator similarity by Brock (1965) and Goethals and Nelson (1973) showed that people are persuaded more easily by communicators viewed as similar, except on matters of verifiable facts. Verification of fact is thought to be more meaningful if the communicator has a different source of information than the receiver. Research by Zajonc (1968) suggests that nothing more than repeated exposure to a person will increase attraction towards that person, even when the first reaction is unfavourable. Other studies also support the notion that people are judged by their appearance. A study recounted by Friend (1985) showed that salesmen were more persuasive on days when they wore coats and ties than when they wore open shirts without ties, presumably because the more formal appearance matches the mental image associated with salesmen.

The attractiveness of the source also seems to enhance persuasion in science and mathematics. A study conducted by Stake and Granger (1978) reported that girls taught by female science teachers were more committed to science careers than girls taught by male teachers. The teachers' influence, according to the researchers, was related to their attractiveness. High science commitment was reported only by those girls who viewed their female science teacher as attractive. Other studies suggest that the persuasiveness of peers and siblings relative to science and mathematics course enrolment and career choice is also a function of attractiveness, with emphasis on the element of source-recipient similarity (Grotevant, 1978; Talton & Simpson, 1985).

As one reflects again on Mr Woods' expression of an attitude favouring gender equity, one must wonder if his attitude is due to his attraction for Ms Jones, the department chair. A gratifying, or perhaps embarrassing, aspect of serving as department chair or in any position of leadership is to observe subordinates express a leader's attitude just because it is that of the leader. As a message source, Ms Jones could influence others through their perception of her attractiveness and their desire to preserve her as part of their own identity.

Source Power

Power is the final category of source characteristics, and it refers to the extent to which the source can control the delivery of rewards and punishments to the recipient. Power works in persuasion, Kelman (1961) writes, by inducing compliance with the position advocated by the source. Attitudes changed by compliance do not reflect private acceptance, but demonstrate a public adoption of the source's position, according to Hass (1981). To produce and maintain compliance, Kelman (1961) and McGuire

(1969) agree that three conditions must be met. First, the message recipient must believe that the communicator has the power to deliver rewards and punishments. Second, the recipient must feel sure that the source will administer the rewards or punishments to induce compliance. Finally, the recipient must believe that she is under the surveillance of the source.

French and Raven (1959) identify five bases from which message sources can derive power: expert, referent, reward, coercive and legitimate. Trenholm (1989) equates French and Raven's expert and referent power bases with the expertness dimension of credibility and attractiveness, respectively. Quite different from other source characteristics discussed thus far, reward power is linked to one's ability to dispense external incentives such as money, grades or promotion. On the other hand, the threat of punishment undergirds coercive power. Employers who dock workers' pay and principals who expel students are exerting coercive power. A key component of both reward and coercion is surveillance. In contrast, legitimate power is derived from societal institutions and does not rely on surveillance for its effectiveness. The right of police officers to maintain order in a disaster area is based on legitimate power.

Parents and, to a much lesser extent, teachers are suspected to function as purveyors of rewards and punishments regarding children and adolescents' science and mathematics choices. Coercive power was found to be a characteristic undergirding the influence of fathers and female science teachers with regard to persuading girls to enrol in high school physical science courses (Koballa, 1988). Scientists also possess power which can affect the persuasiveness of their appeals, but scientists' power is legitimate – stemming from their role as authority figures (Milgram, 1974).

As one thinks for a third time about the conversation between Mr Woods and his science department chair, one might wonder if Mr Woods' expression of a positive attitude towards implementing gender-equity initiatives might have been just for Ms Jones' benefit. He knew that Ms Jones thought that other science teachers should employ teaching practices that promote gender equity and that Ms Jones wanted him to agree. So Mr Woods could have anticipated Ms Jones' favourable reaction to him when he criticised Mr Smith's teaching behaviours that did not foster gender equity. It is also possible that Mr Woods' criticism could have been prompted by his need to avoid an unfavourable evaluation by his department head.

Regrettably, one cannot know exactly which of the source characteristics motivated Mr Woods to voice his opposition to teaching practices that fail to promote gender equity. It would seem reasonable to exclude compliance as the single cause of Mr Woods' behaviour if Ms Jones had heard that he had made similar statements when not in her presence. It is most likely, however, that Mr Woods' attitude was not purely the result of Ms Jones credibility, attractiveness or power. More likely, it came from a blending of all three.

The attitudinal effects produced by the three categories of source characteristics also might not remain distinct, writes Hass (1981). He argues that an attitude produced or changed by identification or compliance later can come to be internalised by a message receiver constructing reasons for holding a new attitude or adopting a new attitude consistent with her actions. Unfortunately, all attitudes acquired by identification or compliance are not internalised. It is only when a new attitude is internalised that its effects on behaviour are prone to be long-lived.

PERSUASION SKILLS

It makes perfect sense, as Trenholm (1989) contends, for a source to stress the personality characteristics that most likely will lead to persuasion. But, what persuasion skills contribute to source credibility, attractiveness and power? And, when will one characteristic be more potent than another? These questions are addressed in this section.

Credibility

There are several ways to establish credibility with respect to issues of gender equity. Expertise is a function of the source's reputation and behaviour. With regard to reputation, past accolades and accomplishments are considered important (Siegel & Sell, 1978). For instance, a person with a record of promoting gender-equity initiatives is likely to be perceived as being more of an expert than a person without such a reputation. When speaking before an audience, evidence of one's expertness can be indicated at the beginning of the speech. Nonverbal behaviours, such as eye contact, attentive posture and context-appropriate gestures, that indicate self-assurance and responsiveness to the audience, often are associated with expertise (Dell & Schmidt, 1976). Use of professional jargon and other verbal cues that supply proof of specialised knowledge also can help to convey expertise (Atkinson & Carskadden, 1975).

Trustworthiness is less dependent on specific behavioural cues than expertise. Research-based suggestions made by Friend (1985) to business managers on how to enhance their trustworthiness seem applicable to a discussion of what can be done to promote gender equity. First of all, Friend recommends the use of straight talk because persons whom you want to persuade most likely will be on the alert for exaggerated claims and extravagant descriptions. Second, he says that the speaker should anticipate objections and raise some of the objections during her speech. Friend also advocates citing specific evidence to support claims, naming information sources, and presenting the message in a calm and candid manner. Moreover, trustworthiness also is associated with the source who appears to argue against her vested interests (Walster et al., 1966) and who is perceived as a recent convert to the position advocated, as in the reformed alcoholic speaking on the evils of drinking (Levine & Valle, 1975).

Attractiveness

There is much that one can do to enhance attractiveness and thus persuasiveness. Attractiveness can be enhanced by conveying liking for the audience and suggesting that the message source and receiver are similar in important ways. Liking can be communicated through nonverbal behaviours, such as head nodding, hand gestures, smiling, voice tone, eye contact and attentive posture (Claiborn, 1979). Use of first-person pronouns and lay jargon also tend to foster liking (Barak *et al.*, 1982).

Speaker-audience similarity can be established by the speaker divulging information about her background, attitudes or values. Similarity seems to promote persuasion when it allows members of the audience to attribute the message to values shared by the source and the audience (Hass, 1981). In this regard, Petty (cited in Davids, 1987) recommends agreeing with the audience on opinions which they hold before presenting them with opinions that they do not hold. While liking and similarity both enhance attractiveness, Kerr *et al.* (1982) contend that, of the two, speaker-audience similarity established verbally can contribute more to the speaker's ability to persuade.

Power

It is generally to the advantage of a source to emphasise power. Often considered synonymous with power are the promise of reward and the threat of punishment. The power of the source enables her to phrase requests to emphasise reward ('If you get an A in chemistry this term, then I will buy you the sports car that you want') or coercion ('Unless you get an A in chemistry this term, you can forget about the new sports car'). The findings of several studies suggest that it is better for a source to use reward rather than coercion to bring about persuasion. One reason is that, when coercive power is used, the message receiver tends to distrust the source, whereas feelings of trust are encouraged when rewards are offered (Kruglanski, 1970). By the same token, the source who uses reward is liked more often than the source who uses coercion (Rubin & Lewicki, 1973). Moreover, a liked source also has the advantage of using other power bases and thereby enhancing and continuing her influence (Raven & Rubin, 1976).

In contrast to reward and coercion, no reason needs to be given by a legitimate source. Legitimate power stems from one person's right to ask another person to comply with a request. The limits of legitimate power are dictated by one's role in a social organisation, such as the family or school. For instance, the teacher, not the student, is the legitimate authority in the classroom. According to Raven and Rubin (1976), words such as 'should', 'ought' and 'oblige' can signal a legitimate power relationship.

Although power can be used effectively to induce compliance, caution should prevail when it is emphasised. Sometimes, according to Trenholm (1989), an excessive show of power, regardless of type, can cause a negative reaction to the perceived loss of freedom. When one believes that she is being controlled, she can come to begrudge this control and renounce the message of the powerful source.

Matching Characteristic and Situation

While it is truly impossible for a source to emphasise characteristics that he or she does not possess, it is possible to analyse an audience, determine which characteristics are needed, and emphasise those characteristics. Based on studies reviewed by Bettinghaus and Cody (1987), Trenholm (1989) advises that credibility should be emphasised when a personal relationship does not exist between the source and the audience, when the proposal proffered is complicated, and when the speaker is indeed an expert. Emphasising attractiveness, she says, is most appropriate when the relationship between the source and receiver is personal and likely to abide, and when the position advocated has to do with personal predilection. Finally, she reasons that power is most effective when the audience has respect for authority, and when rewards and punishments of concern to the audience are controlled by the source.

LEARNING THE SKILLS OF PERSUASION

In learning more about how to foster perceptions of credibility, attractiveness or power, science and mathematics teachers should become aware of the similarities between persuasion and instruction. As Koballa (1992) has pointed out, persuasion and instruction have much in common. Both involve the presentation of arguments and evidence to get someone to believe something or to do something, and both are concerned with the modification and formation of beliefs, attitudes and behaviours that are held with regard to evidence and good reasons. This position not only establishes persuasion and instruction as closely related forms of social influence, but should stimulate discussion regarding the nature of change and how other science-related and mathematics-related beliefs, attitudes and behaviours can be strengthened or modified. It also should foster a broader meaning of and a more positive connotation for persuasion. Science and mathematics teachers will learn that their job as purveyors of knowledge and the desires of society includes governing their own behaviour to bring about intentional effects on the behaviour of students and others in ethical ways.

While schooling in persuasion should focus on the specific behaviours associated with credibility, attractiveness and power, it should not be forgotten that these behaviours most likely will function in tandem rather than in isolation. Therefore, integrating the behaviours into a coherent style of presentation is wise. Individual preferences are likely to become evident as teachers settle on the use of specific cues to project credibility, attractiveness and power. To this end, two teachers can be perceived as equally persuasive while displaying different lexical and behavioural cues.

Projecting an image of credibility, attractiveness or power is only one element of the persuasion process. As tests of the ELM have shown, persuasion skills are no substitute for strong arguments, particularly when the issue is highly or moderately relevant to the receiver. Thus, for issues of high to moderate relevance, the target is more likely to be persuaded when either of the following applies: messages contain arguments developed with regard to compelling evidence and good reasons, or the canons for testing reasons and evidence are offered. According to Crawley and Koballa (1994), beliefs salient to the audience regarding the desired behaviour could be the best source of arguments to include in a persuasive message. As science and mathematics teachers encourage changes in societal goals, such as gender equity, they must come to recognise the importance of both source qualities and message arguments in the persuasion process.

A few prudent cautions should go hand-in-hand with schooling in persuasion. The skills that promote perceptions of credibility, attractiveness and power delineated in this chapter emanate for the most part from research in social psychology, counselling psychology and the fields of marketing and communications, rather than from research in science and mathematics education. The role of source factors in persuasion no doubt will continue to be clarified as research employing the ELM and similar models continues. Thus, the tentative nature of the research-based recommendations made in this chapter always should be on the minds of the teachers of science and mathematics who use them. Schooling science and mathematics teachers in persuasion skills will not change their ethical responsibilities. The use of skills that promote perceptions of credibility, attractiveness or power is ethical when their use contributes to the effectiveness of interventions designed to promote gender equity in science and mathematics. These skills must be used only to achieve ends consistent with the goals of society. Achieving gender equity in the science and mathematics classroom and in the related job market is considered a desirable societal goal. The ethical persuader does not invoke persuasion skills to influence parents, students or others to adopt attitudes or engage in behaviours that are inconsistent with societal goals. The ethical persuader is acutely aware of how the receiver's perceptions enhance her persuasiveness, and uses this power judiciously. Knowledge of the ELM and persuasion skills should equip science and mathematics teachers to function as able and responsible initiators of gender-equity reform.

It is clear that the source of a persuasive message is not the only factor that affects attitude formation and change. But the issues examined in this chapter suggest that the source is important to an understanding of how persuasion can be used to encourage gender equity in science and mathematics. As the research findings presented in this chapter are contemplated and put into practice, the user should be mindful of Bettinghaus' (1968) caution 'that sources can either impede persuasion . . . or help to facilitate it' (p. 120).

ACKNOWLEDGEMENTS

The author wishes to thank Katherine Wieseman for her comments on an earlier draft of this chapter.

The University of Georgia, Athens, Georgia, USA

REFERENCES

Atkinson, D.R. & Carskadden, G.A. (1975). 'A prestigious introduction, psychological jargon and perceived counselor credibility', *Journal of Counseling Psychology* (22), 180–186.

Barak, A., Patkin, J. & Dell, D.M. (1982). 'Effects of certain counselor behaviors on perceived expertness and attractiveness', *Journal of Counseling Psychology* (29), 261–267.

Bettinghaus, E.P. (1968). *Persuasive communication*, New York, Holt, Rinehart & Winston.

Bettinghaus, E.P. & Cody, M.J. (1987). *Persuasive communication*, New York, Holt, Rinehart & Winston.

Brock, T.C. (1965). 'Communicator-recipient similarity and decision change', *Journal of Personality and Social Psychology* (1), 650–654.

Byrne, D. (1971). *The attraction paradigm*, New York, Academic Press.

Chaiken, S. (1979). 'Communicator physical attractiveness and persuasion', *Journal of Personality and Social Psychology* (37), 1387–1397.

Claiborn, C.D. (1979). 'Counselor verbal intervention, nonverbal behavior, and social power', *Journal of Counseling Psychology* (26), 378–383.

Crawley, F.E. & Koballa, T.R. (1994). 'Attitude research in science education: Contemporary models and methods', *Science Education* (78), 35–55.

Davids, M. (1987). 'Believe me', *Public Relations Journal* (43), 16–18 & 43.

Dell, D.M. & Schmidt, L.D. (1976). 'Behavioral cues to counselor expertness', *Journal of Counseling Psychology* (23), 197–201.

French, J.R. & Raven, B.H. (1959). 'The bases of social power', in D. Cartwright (ed.), *Studies in social power*, Ann Arbor, MI, University of Michigan Press, 150–167.

Friend, W. (1985). 'Winning techniques of great persuaders', *Association Management* (37), 82–86.

Goethals, G.R. & Nelson, R.E. (1973). 'Similarity in the influence process: The belief-value distinction', *Journal of Personality and Social Psychology* (25), 117–122.

Grotevant, H.D. (1978). 'Sibling constellation and sex typing of interest in adolescents', *Child Development* (49), 540–542.

Hass, R.G. (1981). 'Effects of source characteristics on the cognitive processing of persuasive messages and attitude change', in R. Petty, T. Ostrom and T. Brock (eds.), *Cognitive responses in persuasion*, Hillsdale, NJ, Erlbaum, 141–172.

Hass, R.G. & Grady, K. (1975). 'Temporal delay, type of forewarning and resistance to influence', *Journal of Experimental Social Psychology* (11), 459–469.

Heesacker, M., Petty, R.E. & Cacioppo, J.T. (1983). 'Field-dependence and attitude change: Source credibility can alter persuasion by affecting message-relevant thinking', *Journal of Personality* (51), 653–666.

Hensel, R.A.M. (1989). 'Mathematical achievement: Equating the sexes', *School Science and Mathematics* (89), 646–653.

Hovland, C.L., Janis, I.L. & Kelley, H.H. (1953). *Communication and persuasion*, New Haven, CN, Yale University Press.

Kahle, J. & Meece, J. (1994). 'Research on gender issues in the classroom', in D. Gabel (ed.), *Handbook of research in science teaching and learning*, New York, Macmillan, 542–557.

Kelman, H.C. (1961). 'Process of opinion change', *Public Opinion Quarterly* (25), 57–78.

Kerr, B.A., Claiborn, C.D., Dixon, D.N. (1982). 'Training counselors in persuasion', *Counselor Education and Supervision* (22), 138–148.

Koballa, T.R. (1988). 'Persuading girls to take elective physical science courses in high school: Who are the credible communicators?' *Journal of Research in Science Teaching* (25), 465–478.

Koballa, T.R. (1992). 'Persuasion and attitude change in science education', *Journal of Research in Science Teaching* (29), 63–80.

Kruglanski, A.W. (1970). 'Attributing trustworthiness in supervisor-worker relations', *Journal of Experimental Social Psychology* (6), 214–232.

Levine, J.M. & Valle, R.S. (1975). 'The convert as a credible communicator', *Social Behavior and Personality* (3), 81–90.

Martin, R.E. (1985). 'Is the credibility principle a model for changing science attitudes?', *Science Education* (69), 229–239.

McGuire, W.J. (1969). 'The nature of attitude and attitude change', in G. Lindzey and E. Aronson (eds.), *The handbook of social psychology*, Vol. 3 (second edition), Reading, MA, Addison Wesley, 136–314.

Milgram, S. (1974). *Obedience to authority*, New York, Harper & Row.

Miller, R.L., Brickman, P. & Bolen, D. (1975). 'Attribution versus persuasion as a means for modifying behavior', *Journal of Personality and Social Psychology* (31), 430–441.

Petty, R.E. & Cacioppo, J.T. (1981). *Attitudes and persuasion: Classic and contemporary approaches*, Dubuque, IA, W.C. Brown.

Petty, R.E. & Cacioppo, J.T. (1986). *Communication and persuasion*, New York, Springer-Verlag.

Petty, R.E., Cacioppo, J.T. & Schumann, D. (1983). 'Central and peripheral routes to advertising effectiveness: The moderating role of involvement', *Journal of Consumer Research* (10), 135–146.

Puckett, J., Petty, R.E., Cacioppo, J.T. & Fisher, D. (1983). 'The relative impact of age and attractiveness stereotypes on persuasion', *Journal of Gerontology* (38), 340–343.

Raven, B.H. & Rubin, J.Z. (1976). *Social psychology: People in groups*, New York, John Wiley.

Rhine, R.J. & Severance, L.J. (1970). 'Ego-involvement, discrepancy, source credibility and attitude change', *Journal of Personality and Social Psychology* (16), 175–190.

Rubin, J.Z. & Lewicki, R.J. (1973). 'A three-factor experimental analysis of interpersonal influence', *Journal of Applied Social Psychology* (3), 240–257.

Rudisill, E.M. & Morrison, L.J. (1989). 'Sex differences in mathematics achievement: Emerging case for physiological factors', *School Science and Mathematics* (89), 571–577.

Saergert, S., Swap, W.C. & Zajonc, R.B. (1973). 'Exposure, context and interpersonal attraction', *Journal of Personality and Social Psychology* (25), 234–242.

Shrigley, R.L. (1976). 'Credibility of the elementary science methods instructor as perceived by students: A model for attitude modification', *Journal of Research in Science Teaching* (13), 449–453.

Shrigley, R.L. (1980). 'Science supervisor characteristics that influence their credibility with elementary school teachers', *Journal of Research in Science Teaching* (17), 161–166.

Siegel, G.C. & Sell, J.M. (1978). 'Effects of objective evidence of expertness and nonverbal behavior in client-perceived expertness', *Journal of Counseling Psychology* (25), 188–192.

Stake, J.E. & Granger, C.R. (1978). 'Same-sex and opposite-sex teacher models as influences on science career commitment among high school students', *Journal of Educational Psychology* (70), 180–186.

Talton, E.L. & Simpson, R.D. (1985). 'Relationship between peer and individual attitudes toward science among adolescent students', *Science Education* (69), 19–24.

Trenholm, S. (1989). *Persuasion and social influence*, Englewood Cliffs, NJ, Prentice Hall.

Trigg, L.J. & Perlman, D. (1976). 'Social influences on women's pursuits of a nontraditional career', *Psychology of Women Quarterly* (1), 138–150.

Walster, E., Aronson, E. & Abrahams, D. (1966). 'On increasing the persuasiveness of a low prestige communicator', *Journal of Experimental Social Psychology* (2), 325–342.

Zajonc, R.B. (1968). 'Attitudinal effects of mere exposure', *Journal of Personality and Social Psychology Monograph Supplement* (9), 1–27.

DORIS JORDE[1] AND ANNE LEA[2]

13. SHARING SCIENCE:
PRIMARY SCIENCE FOR BOTH TEACHERS AND PUPILS

The following is a story about two primary school teachers, Sara and Elizabeth, and their science teaching experiences. Sara and Elizabeth have the same educational background including attending a teacher training school after completion of high school. Both have concentrated on languages at the high school level and later in their education courses. Both have four years' experience and currently are teaching grade 5. They teach in the same school district but in two different schools.

Their local school district recently has written a detailed plan of what topics should be covered at each level of the primary school. What follows is a description of how Sara and Elizabeth interpreted the science section of that plan, particularly the topic of sound.

PLANNING A UNIT ON SOUND: TWO INTERPRETATIONS

Planning

Sara read the science plan for the coming year with little enthusiasm. She still had her doubts about teaching science to children because she had not taken much science herself. In fact, the last science class that Sara had taken was when she was 16 years old during her first year of high school.

However, there it was in writing: the topics to be covered in grade 5 included flowering plants, weather, the human body, light and sound. Luckily, the science textbook which Sara had chosen for her fifth graders included an entire chapter on sound. Sara decided to follow the text and let the children try some of the experiments suggested in the book.

Elizabeth read the science plan for the coming year with a good deal of anticipation. She too had received no formal training in science since high school. Luckily, however, she had recently picked up lots of ideas when she attended an inservice course devoted to science topics for young children. This would be her first chance to try out some of these new activities with her class.

Elizabeth used the children's textbook to give her background information for the unit. She was quite pleased to see that the book included a number of experiments to which she could add those which she had learned at the inservice course. She began designing her unit on sound, basing her design on a progression of experiments that would help the children to discover some basic concepts about sound.

155

L.H. Parker et al. (eds.) Gender, Science and Mathematics, 155–166.
© 1996 *Kluwer Academic Publishers. Printed in the Netherlands.*

Opening

Sara started her unit on sound by asking the general question, 'What is sound?' She then took a rubber band and began plucking it so that the children could hear a sound. She asked the class, 'What happens when I pluck the rubber band?' Many children had their hands up in the air but Lars shouted out the answer: 'The rubber band vibrates and makes a sound'.

A series of questions and answers followed her demonstration. Sara led the discussion on sound so that the information which she thought was important to learn would be discussed. After one class period, her class had discussed the ideas that: sound comes from things that vibrate; and sound waves travel through the air and into our ears so that we hear.

After her demonstration, Sara asked the class to read the appropriate pages in their science book and to try some of the experiments at home with their parents. She indicated that, later in the week, everyone would be asked to demonstrate one of the experiments that they had tried at home.

Elizabeth began her unit on sound by asking everyone to write everything that they knew about sound. She gave the message that the compositions would not be corrected. The purpose of the writing assignment was for students to find out just how much they already knew about sound and later to share their ideas with one or two other children. As Elizabeth's class had used this writing technique many times before in their language lessons, they knew that they should write whatever came to their minds. They also knew that they had a limited amount of time to write, so they began at once to record their thoughts.

After 15 minutes, everyone was asked to stop writing and to move their desks together with a neighbour. The assignment then was to read their compositions to each other and to talk about their ideas. Elizabeth gave them one more rule: 'You may only make positive comments about each other's ideas!'

The final set of directions was for each group to write two or three questions about things that they wondered about sound. Elizabeth collected these questions and hung them on the bulletin board.

After the writing assignment was completed, Elizabeth gave everyone a rubber band. She asked them to make as many different sounds as possible with their rubber bands. After about 10 minutes of exploration time, she assembled the children to see how many different sounds the children were able to make.

Progression

Sara completed her unit on sound in one week. After her opening demonstration and discussion, she asked the children to read about sound in their textbooks. She conducted additional discussions about sound during the week, when topics from the book were discussed. Sara was always the leader of these discussions. The children discussed the ear and its parts, the speed of sound, echoes, high and low sounds, loud and soft sounds, and musical instruments. The high point of the week was when the children presented an experiment on sound that they had tried at home.

Elizabeth completed her unit on sound in two weeks. After the writing assignment and experimenting with the rubber bands, she decided to expand on the idea that sound is produced by vibrating objects. A series of activities followed which gave the children a chance to explore this concept. The children made their own musical instruments. In addition, they demonstrated the instruments which they played in the school band and they experimented with tuning forks.

Elizabeth's class also discussed the ear and its parts. They experimented with the shape of ears by making bigger ears for themselves out of paper. They cut out pictures of animal ears from magazines, placing them into groups according to their shape and size. They learned the letters of the alphabet using sign language.

Elizabeth's final set of activities demonstrated the concept that sound travels through different mediums. Her class made telephones with cans and string. They listened to sounds travelling through their desk. They listened to sounds under water. They talked about sound on the moon.

Elizabeth completed her unit by having her class write once again about sound. She then returned their original compositions to them and encouraged them to compare what they knew before with what they knew now.

Sara or Elizabeth – Which Example Do We Prefer?

We know that Sara exemplifies a typical primary teacher when it comes to teaching science topics. She is intimidated by science because of her educational background, yet she knows that she must and should teach science topics. She is willing to give science a try, but has only her textbook to guide her. Indeed, experiments are a part of her unit on sound, but they are teacher directed and do not follow a clear line of concept development.

Elizabeth's unit on sound is an example of what we consider to be good primary science. It is good because it takes into consideration the fact that children are active learners who need to construct their own meaning through experiences with the world. It is good because it challenges children to explore and discover, yet also guides them in their discovery. It is good because all children, both girls and boys, are involved in the lessons. It is good because Elizabeth sees her role as an instigator and learns together with her students.

All teachers like Sara and Elizabeth need help in planning and designing science units. The Primary Science Project in Norway was developed in order to provide a 'user friendly' science model for primary teachers. When we look more closely at Elizabeth's unit on sound, we will begin to discover what makes a successful primary science program.

THE PRIMARY SCIENCE PROJECT

When the Primary Science Project began in 1986, there was a good deal of statistical information available on the state of primary science in Norway. We participated in the Second International Science Study (Sjøberg, 1986) and had been involved in a large project involving girls and physics (Lie & Sjøberg, 1984).

We knew that primary science was not considered a priority in the curriculum and that teachers had little or no formal science education. We also knew that science had been integrated into a subject called 'orienteering', although we knew little about how this new subject was being taught. Finally, we knew that schools spent little or no money on science equipment.

The Second International Science Study gave us information regarding children's attitudes, achievement and experiences in science. It showed us that, generally, grade 4 pupils had a positive attitude towards science. What was alarming was the general decline in positive attitudes between grade 4 and grade 9. This was especially apparent with girls! Gender differences also became apparent in achievement test scores, with boys generally performing better than girls.

Data on experience with science objects and/or topics outside the classroom gave very interesting information on the differences between what boys and girls experience with informal science. Boys had much more experience and interest with physics topics (model building, using tools, repairing things, changing batteries, etc.). Girls had experience and interest in biology topics (plants, human body, animals, food, etc.).

After a careful assessment of what we thought that we knew about primary science, it became apparent to us that there were many pieces of information of a qualitative nature that were lacking, especially in relation to actual classroom teaching situations. We had important questions that needed to be answered concerning gender issues in science teaching, the methods used to teach science topics, the content knowledge required, and the combinations of methods and content found.

We came to the project with a constructivist perspective on what we considered to be good science teaching (Driver, 1983, 1989). In our view, children are responsible for what they learn, they come to the science classroom with notions of the world around them, and they need to be challenged to explore their own alternative conceptions before being willing to accept new and/or different concepts. We were convinced that an activity-based science curriculum was the correct way to teach primary science, based on what we knew about conceptual change learning.

With these thoughts in mind, we began visiting primary classrooms to observe science teaching. What a great disappointment it was for us to discover that there was literally no science teaching to observe. The closest that we came to a science lesson was a reading assignment in a textbook.

CURRICULUM DEVELOPMENT

We set out to create an activity-based science curriculum that could be used by our primary teachers. We worked together with 12 grade 4–6 teachers from two schools throughout the curriculum development process. The teachers used the materials in their classes while we observed them in action. Their comments, together with our observations and interviews, served as the basis for the design and evaluation of the materials.

Teachers as Researchers

We cannot say enough about the importance of using classroom teachers in the development of curriculum materials. Indeed, we came into the project with a perspective on how children can best learn science. But, it was the teachers using the materials who were able to evaluate the methods of presentation which we recommended, the level of language used in the text, the time needed to complete the activities and the degree of difficulty for each of the activities.

We met with teachers before they taught the new materials to give them an overview of what we expected in the activities. We also had casual conversations with the teachers while they were involved in teaching the science topics. But the most helpful and interesting teacher involvement came in our group interviews after all of the teachers had completed the units. This was the time when they talked to each other about their experiences and were able to tell what worked and what did not work. This was also the time when they expressed their own doubts about how they presented lessons, especially when the science content was beyond what they knew.

Magnets, Mystery Powders, Sound and Light

In a two-year period, we developed four units entitled *Magnets, Mystery Powders, Sound and Light*. In each of the units, we tried to incorporate new methods for teaching science, always trying to improve on our model of curriculum development.

In our first unit *Magnets* (Lea *et al.*, 1989) we were concerned with developing science activities that would involve all pupils actively. We concentrated on methods for opening a unit that could attract the attention of all pupils, and not just those who had experimented with magnets before. We learned the importance of using activities to give all pupils a common ground before going further in their discoveries.

In the second unit *Mystery Powders* (Jorde *et al.*, 1989), we were looking at how children work in groups. The activities and the reporting were all group efforts. In addition, we looked at the role of the teacher during group activities. We learned how important it is for teachers to have methods of interrupting group work in order to guide the lesson and keep the groups going in the right direction. Teachers successful in managing group work also were teachers who had a means of showing pupils that they were in control of the class.

The unit *Sound* (Jorde, 1989) was used to experiment with the use of descriptive writing in the science lesson; this unit is discussed in the latter part of this chapter. The fourth unit *Light* concentrated on uncovering children's alternative conceptions about light.

After extensive testing and evaluation in classrooms, we developed a working model which guided the structure of the individual units. We combined our knowledge of the primary classroom together with research on primary science teaching and learning. In developing our materials, we experimented with the fine balance between content, process and method, a balance which we learned to appreciate through classroom teachers.

The materials consist of two parts. First, a student workbook includes the activities for the children and a textbook. The activities are meant to be carried out in the classroom under the guidance of the teacher. The textbook can be read in the classroom or used for homework. The textbook is used to present scientific theory, provide explanations for the activities carried out in the classroom, and information which places the science topic into a broader societal context. Second, a teacher's guide includes all of the information in the student workbooks together with detailed information on each activity. A description of the activity, examples of typical questions which children are likely to ask, and a little theory on the science topic are included for each activity.

SWINGING STRINGS – A UNIT ON SOUND

What follows is a description of what we developed and learned while working with our unit *Sound*. As mentioned before, we tried to experiment with new methods each time we began a new unit. We used the sound unit to experiment with writing in the science lesson. We were interested in using writing to get at what children already knew about sound, and to see if writing encouraged greater participation in the lessons and led to improved retention of science concepts.

Our content objectives for the unit were simple. We wanted children to understand the ideas that sound comes from things that vibrate, that sound travels and that we use our ears to hear sounds. The progression of activities found in the unit on *Sound* is shown in Figure 1.

Writing about Sound

The unit on sound was intended to emphasise the use of writing in the science lesson. Children wrote about sound before the activities began (pre-instruction), at the end of the unit on sound (post-instruction) and one year after completion of the unit.

The pre-instruction writing assignment was used to learn about the ideas which children had about the topic. The assignment was not given in the context of the science lesson and reflected, therefore, the children's general thoughts about sound. In almost every case, children from grades 4–6 chose to write about types of sounds,

1. What do we know about sound? Children write everything that they can about the topic of sound. They later read what they have written for each other and write two questions on things they wonder about with sound. Every child then is given a rubber band with which to experiment. The homework assignment is to make a musical instrument.

2. Sound comes from things that vibrate. Children present instruments which they have made at home. They work in groups to make a song or poem using their instruments to accompany them. After presentations to the class, each instrument is observed to determine what it is that actually makes the sound.

3. Swinging Strings. We provide a variety of ways in which to experiment with sounds created by plucking strings. We introduce the concept of pitch and loudness.

4. Is there sound in outer space? A number of activities that demonstrate how sound travels through different media are provided. Children then make a telephone using yoghurt cups and string, followed by experiments using their telephones.

5. Sound takes time. This activity involves measuring the speed of sound out in the playground as well as finding an echo wall.

6. How do we hear? We study our own ears in this activity, especially how sound waves travel from a source into our ears and eventually end up in a message to our brain. The shape of ears is discussed with reference to animal ears.

7. What do we use sound for? This activity introduces sound as a part of our environment. The sign alphabet for deaf people is introduced.

At the end of the unit, pupils once again were asked to write about sound as they had done in the first exercise.

Figure 1.   Progression of activities in the unit *Sound*.

using lots of adjectives to describe sounds. Scientific terms were not found in the majority of these essays, apart from isolated cases for which sound waves were mentioned. If a scientific term was mentioned, it was a boy who wrote the essay.

The information obtained from these first assignments told us that children in this age group do have a concept of sound which is connected to sounds that they hear. They did not chose to write about sound as we might describe it, in a scientific way.

The purpose of the post-instruction writing assignment was to provide information on what was learned and retained by the children. When compared with the pre-instruction assignments, children were more able to write about sound using scientific terms. Often the assignments reflected upon the activities performed in the unit.

The information obtained from the second writing was used to evaluate our curriculum materials. We found that pupils had retained many of the simple concepts demonstrated through the experiments and reading assignments. We also were able to identify instances in which children had formed alternative concepts about sound based on the activities in the unit. This information was very important in the curriculum evaluation and revision process. We tried to establish whether the 'misinformation' had come from the activities, from the teacher explanations, or from a combination of the child's previous ideas about sound and new information.

After one year, scientific terms and concepts about sound were used in the writing assignments. Our impressions when reading the essays were that concepts were related directly to experiments which children remembered doing. Many students even described the experiments which they remembered. The importance of activity-based science cannot be underestimated when we see how the activity is coupled with retention.

'GOOD' PRIMARY SCIENCE

In 1990, an article in *Newsweek* entitled 'Rx for Learning' discussed science education in the USA. It stated that the only issues in science education are deciding what to teach and then deciding how to teach it. One could say that this is the essence of the field that we call 'science education'. As we have studied primary science over the past years, we have discovered that these two issues are woven intimately together in every science lesson. It is not enough to have content if one does not have any idea how to motivate children so that they will want to learn. On the other hand, it is of little use to practice new and exciting classroom methods if one does not have some content to teach.

Our struggle to define the interaction between content and method, between theory and practice, has led us to a description of what we believe constitutes good primary science. The most important messages that can be transferred from our project to the reader are the essential components which we have identified below for making a successful primary science curriculum.

1. *All children must be motivated and engaged from the very first introductory remark.* How many times have we seen and heard teachers introduce a new theme by asking the entire class a question like: 'What do you know about magnets?' Anyone working in the classroom can predict the outcome of such an opening. Only those students who know what a magnet is will raise their hands, and usually a few boisterous boys will shout out the answer before anyone else has a chance to think. We do not recommend such an opening if we are interested in gender-inclusive science. In our units, we experimented with introductions or openings to lessons that will attract the attention of all students, regardless of their previous experience or lack of experience with science topics.

Whether it is an activity or a writing assignment, the goal of the introduction or the opening is to allow all children the opportunity to know that they have something to contribute, and that they have an interest in the topic. Motivation to learn more is the key ingredient.

2. *Science learning requires structure and guidance as well as activities.* Activities alone are not enough for good science teaching. We cannot expect that children will 'discover' the principles of science on their own, as the great scientists of the past have done. Driver (1983) has turned the often used proverb of 'I hear, and I forget; I see, and I remember; I do, and I understand' into: *I do, and I am even more confused!*

A child who has had the opportunity to 'play' with a magnet has little chance of arriving at the same level of understanding that has developed about magnets over hundreds of years. The activity plays the important role of motivating the child into wanting to know more about magnets and increasing the possibility for the child to understand scientific knowledge about magnets. However, discovery learning without guidance in learning about the principles and theories of science could lead to play and confusion.

So, once we have the attention and interest of the children, it is time to get to work on learning some science. As stated before, an activity-based science curriculum is essential if we are to take our views on conceptual change learning seriously. However, we do not believe that the road to learning is through activities alone. The teacher must be willing to act as a guide and/or leader throughout the activity, so that children are aware of what they should be observing, what they should be doing, and what they should be thinking about.

3. *An exciting science curriculum has many different presentation forms.* We thought that, because elementary teachers have a strong concentration on pedagogy, this would be reflected in science teaching through the use of varied methods of presentation. This was not the case. Teachers who were very clever in using group work, drawing and oral presentations in other subjects sometimes were left helpless when asked to teach a science lesson. Guidelines for group work (see Johnson & Johnson, 1987) suggest that each activity must be evaluated according to learning objectives and that decisions must be made about whether the children will be working alone on the task, in groups of two or in larger groups. Like many others working on grouping in science, we have observed that groups of 3–4 children are generally large enough.

Gender-inclusive science should have a strong writing component. There are many children who are strong in writing, yet have problems with oral communication. By writing, we mean more than scientific reporting through journal writing. Many of the writing assignments with which we have experimented are of an expository nature, involving children in writing freely about what they think or what they have experienced. Sometimes the assignments involve writing to a visitor from outer space, describing what has been going on. At other times, they involve writing about the science lesson to a sick classmate or to a parent. The important point with writing is that children are forced to organise their thoughts about what they have done and communicate them through writing. The McClintock Collective in Australia (Gianello, 1988) has been very successful with the use of writing in science lessons and has inspired our work as well.

Drawing is an important form of expression for younger children, but it can be a surprisingly rich form of communication for older children as well. Often, group work is summarised best by making a poster which demonstrates what has been done. These posters have the added attraction of decorating the often blank walls of the classroom with children's work in science.

The Norwegian people have a very strong connection to nature and the outdoors. It is not difficult, therefore, to suggest that science activities take place outside when appropriate. Norwegian children always have the correct clothes for going outside in any weather, including rain and snow. For example, once the children undertaking our unit had made compasses on magnets, they went outside and used their compasses to guide them while their heads were covered by large paper bags. Our unit on sound had children outside to listen to sounds and experiment with the speed of sound and echoes.

When you walk into a classroom where children's work is presented on the bulletin boards, where colourful posters decorate the walls, where equipment is out to be used rather than stored in cupboards, where books are readily available for extra reading, you immediately have the feeling that this is a classroom where children want to be. Research by Jane Butler Kahle (1987) in the USA has indicated that visually stimulating classrooms are an important factor in encouraging girls in secondary science. We feel that this applies for the elementary science classroom as well.

4. *Less content is more.* In Norwegian elementary schools, curriculum is not driven by assessment. There is no national assessment of any kind in any subject at this level of instruction. We are able to adopt the phrase 'less is more' as a reality for primary science. Fewer content areas, given more time, mean a more interesting and challenging science curriculum for the elementary level.

Primary science should not be accused of being a 'watered-down' secondary science, full of difficult words and even more difficult concepts. We argue that primary science should not be filled with facts to be memorised and difficult concepts which are not understood. Rather, it should be a science that combines content with the processes of science and allows children to discover things by using their senses, ask questions, be inquisitive, find relationships, make predictions based on their observations, create new things and communicate what they are doing.

5. *Teachers are also learners.* Primary teachers in Norway are no different from their counterparts in other western countries: they lack a formal science education and therefore lack confidence in teaching science. The teachers with whom we have worked have been very honest about their lack of science knowledge. This was not a big problem when we introduced the units *Magnets* or *Mystery Powders*. But, teachers immediately reacted in a negative way to the *Sound* unit. Sound, after all, is physics and has something to do with waves and other incomprehensible things.

Curriculum materials must have a well-developed teacher's guide that gives teachers step-by-step information on how each activity should look. This includes information on time, methods of presentation, typical questions which children might ask, answers to some of those questions, and last but not least, content information which empowers teachers. If curriculum guides for teachers do not concentrate on activities, then the activities are not done and we are back to science as a subject which one reads rather than a subject which one does.

6. *Total school involvement leads to increased enthusiasm and participation.* At the school level, the ingredients for a successful science program at the primary level include: a strong administration willing to support new ideas and programs by allocating both time and money to science; groups of teachers trying out the same activities and sharing their experiences with each other; common equipment for all teachers, organised according to themes; inservice courses in primary science which involve at least two teachers from the same school participating and reporting back to colleagues.

GENDER-INCLUSIVE SCIENCE – A SCIENCE FOR ALL CHILDREN

During the Gender and Science and Technology (GASAT) conference in Sweden in 1990, one of the authors of this chapter was inspired to write the following verse during a writing assignment:

> Gender-Inclusive Science is a science made for all.
> We emphasise activities and not just cold recall.
> It's breaking out of old ideas: away from what's been done,
> So kids can see and do and learn and say that
> Science is fun!

The verse describes an important point about gender-inclusive science: when special consideration is given to providing a science curriculum that is more suitable for girls, the curriculum is better for all children. Girls and boys benefit from a science curriculum that includes pictures of both sexes in the textbooks (for example, using toasters as well as cars), individual activities for which everyone has something to contribute, and group activities for which everyone has an important task to perform.

Discussions of gender-inclusive science, however, rarely involve the importance of improving science for teachers. In most western countries, the primary teacher workforce is made up of women. We know that girls avoid science in school, and often become primary teachers who perpetuate a negative spiral for science. They often neglect teaching science topics in the primary curriculum because they lack self-confidence.

The problems which female teachers face in teaching primary science have been of utmost importance to the project described in this chapter. No matter how exciting activity-based science is for children, if teachers are unable and unwilling to try such curriculum materials in their own classrooms, then gender-inclusive science will not become a reality. The needs of the teacher are equally as important to a gender-inclusive science program as those of the children.

¹University of Oslo, Norway;
²College of Early Childhood Education, Oslo, Norway

REFERENCES

Driver, R. (1983). *The pupil as scientist?* Milton Keynes, Open University Press.
Driver, R. (1989). 'Changing conceptions', in P. Adey (ed.), *Adolescent development and school science*, London, Falmer Press, 79–103.
Gianello, L. (ed.) (1988). *Getting into gear – gender inclusive teaching strategies in science*, Canberra, Curriculum Development Centre.
Johnson, D.W. & Johnson, R.T. (1987). *Learning together and alone: Cooperative, competitive and individualistic learning* (second edition), Englewood Cliffs, NJ, Prentice-Hall.
Jorde, D. (1989). *Sound in the elementary curriculum*, paper presented at the Nordic Conference on Science and Technology Education, Heinola, Finland.
Jorde, D., Lea, A., Baalsrud, K. & Hannisdal, M. (1989). *Spennende stoffer* (Mystery powders), Oslo, H. Aschehoug & Co.
Kahle, J.B. (1987). 'Teachers and students: Gender differences in science classrooms', in J.Z. Daniels and J.B. Kahle (eds.), *Contributions to the Fourth GASAT Conference* Vol. 3, Ann Arbor, MI, University of Michigan, 18–25.
Lea, A. (1989). 'Aktivitetsbasert naturfag påbarnetrinnet', Unpublished dissertation, Norway, Center for Science Education, University of Oslo.
Lea, A., Jorde, D., Hannisdal, M. & Larsen A. (1989). *Merkelige magneter* (Magnificent magnets), Oslo, H. Aschehoug & Co.
Lie, S. & Sjøberg, S. (1984). *'Myke' jenter i 'harde' fag?* (Soft girls in hard science?), Oslo, Universitetsforlaget.
Sjøberg, S. (1986). *Naturfag og Norsk skole* (Science in Norwegian schools), The National Version of the Second International Science Study, Oslo, Universitetsforlaget.

14. 'DO YOU KNOW ANYONE WHO BUILDS SKYSCRAPERS?'

SOS – SKILLS AND OPPORTUNITIES IN SCIENCE FOR GIRLS

I didn't know that architecture was anything to do with science!
I thought engineering was about cars and then Kim told us about her job and it was really good!

For two days in November 1989, 70 form 4 girls (aged 13–14 years) from Auckland Girls' Grammar School in New Zealand were faced with a variety of problem-solving activities directly related to industry. They were led through these activities by women scientists working in industry. From the moment when our first role model, Kim Rutter, introduced herself as 'someone who builds skyscrapers' to Anske Jannssen's last frenetic marketing exercise, the pace was cracking.

SOS – Skills and Opportunities in Science for Girls is an intervention program developed to give young women a glimpse of industrial work through the eyes of role models. Recognising that, all too often, young people are 'talked at' when the merits of particular careers and subject choices are explained, the SOS team, consisting of Liz Godfrey (University of Auckland), Lorraine McCowan (Auckland College of Education) and myself, decided to develop a program that involved the young people participating and kept educators in the background. I had been involved in the Careers Research and Advisory Centre (CRAC) 'Education for Industry' program in Cambridge, UK and hoped that a modified program could demonstrate that careers in industry were not only feasible for women, but also an exciting possibility.

Once Liz and Lorraine joined me, the concept took off. We developed a workpack and videotape to explain the objectives of the program and serve in the training of prospective schools and role models. Specifically, the workpack lists the objectives of the program as:
• to show girls the varied career opportunities available when they have studied science at senior levels;
• to illustrate the work of scientists and engineers;
• to give practical experience of the skills used by scientists and engineers through problem-solving exercises.

Since the first SOS conference held at Auckland Girls' Grammar School in 1989, we have organised conferences throughout New Zealand and developed a range of problem-solving activities that relate directly to the jobs of our role models. Below are described some typical problem-solving activities which we have used.

Eggstraordinary involved a mechanical engineer in demonstrating the qualities, skills and challenges involved in her job. The group is given the task of making, out of LEGO Technic, an eggbeater which is able to beat an egg for one minute. The vocal support for this activity during a tension-filled beat-off raises the roof and we marvel at the range of solutions offered.

167

L.H. Parker et al. (eds.) Gender, Science and Mathematics, 167–176.
© 1996 *Kluwer Academic Publishers. Printed in the Netherlands.*

The Choicest Jelly enabled a food technologist to share her experiences in developing and promoting a new food product. The groups are presented with a less than appetising jelly and asked to improve the product, find a market and sell their promotional strategy to their peers. The role model provided a glimpse into the food-processing industry where problem-solving inevitably is linked with the subtleties of marketing.

The Great Park Experience allowed an environmental engineer to demonstrate that engineers must take account of geology, topography and the community's wishes, as well as keep within strictly defined budgets, when developing a recreational area. Numerous imaginative models were developed and a number of inspiring philosophies were proposed during this activity.

It would be presumptuous to claim that this intervention program is the solution to the under-representation of girls in the physical sciences. Instead, I will provide some details of the problem of low participation rates of girls in physical sciences in New Zealand, explain the SOS philosophy, describe SOS as an example of an intervention program, quote some comments from our first SOS participants, discuss the limitations of the program and share some of our plans for the future.

IS THERE A NEED FOR SOS?

The low uptake of physical sciences by women at universities is well documented (e.g., Brickhouse *et al.*, 1990; Byrne, 1993). Indeed, by the turn of the century, there could be such a shortfall of engineers that the major industrial economies could be competing for their services. Thus, many less affluent countries are likely to be priced out of the market and face technological and economic stagnation (O'Neill, 1990). It is inevitable that the massive shortage of scientists and technologists in North America, Western Europe and Australia is likely to precipitate a brain drain from New Zealand (Geddes, 1991).

Also, in common with many other countries, there are very few women graduating in the physical sciences and engineering. The appointment in 1989 of Liz Godfrey as Liaison Officer for Women in Physical Sciences and Engineering is evidence that Auckland University is aware of the seriousness of the problem. Liz's major role is to increase the intake of women into the physical sciences and engineering.

Whilst it is encouraging that the number of female students staying on for form 7 (16–17 years) has trebled from 1980 to 1991 – from 13.4% to 42% of the form 3 (12–13 years) pool (Sturrock, 1993), it is disappointing that the proportion of these women students engaged in physical sciences at form 7 remains low (see Table I).

This low uptake could be related to the wider range of subjects available to form 6 and 7 students or it could be related to the perceived usefulness of these subjects to their future careers. Project FAST – Future Aspirations, Subjects and Training of Senior Secondary School Students (Rivers *et al.*, 1989) identified and analysed a sample (14,000) of form 7 students' subject and occupational preferences. Table II demonstrates the low ranking of science-based occupations in their choices.

TABLE I

Percentage of male and female students taking form 7 Biology, Chemistry and Physics in 1990[a]

	% of students choosing science in senior school	
Subject	Female	Male
Biology	42.9	30.3
Chemistry	23.1	31.3
Physics	14.5	40.7

[a]From Ministry of Education (1991).

TABLE II

New Zealand form 7 students: Ranked preference of science-based careers of 50 career options[a]

	Rank of preferred careers (1–50)	
Science based careers	Female	Male
Engineer (professional)	39	2
Doctor	5	11
Scientist	31	20

[a]From Rivers *et al.* (1989).

PHILOSOPHY OF SOS

A review of the international literature reveals many intervention programs designed to persuade young women to enter careers in the physical sciences (e.g., Girls Into Science and Technology (GIST), Women Into Science and Engineering (WISE), Girls and Technical Education (GATE), Girls and Maths and Science Teaching (GAMAST), Participation and Equity Program (PEP) and EQUALS=Science). These range from school-based programs involving the development of 'girl-friendly' curricula, to ongoing enrichment programs and one-off consciousness-raising sessions that can vary in duration from a few hours to several days, and to programs directed at teachers' preservice or inservice education.

Economic constraints meant that, in SOS, we had to focus on a short intensive program that presented 'our' view of science to young women. In summary, our hidden objectives were to provide a program of problem-solving experiences for which cooperation, creativity, intuition and communication skills were recognised in an atmosphere of mutual respect. Also, we believed that it was important to spend money and time creating a warm, attractive conference atmosphere for the participants.

What makes SOS such a popular package? What is it that students enjoy? A quote from an observer at the conference provides a practical answer to these questions:

I came away from two days of observing . . . feeling incredibly 'buzzy' – and the fourth formers who had attended gave every impression of feeling the same. From start to finish, they were on the go –

designing, creating, and making a variety of off-beat products . . . as they tested and experienced the excitement of personal discovery.

And women were in command! Each activity was facilitated by an articulate involved woman working in science, who had her own slant on science in the 'real world' (e.g., an engineer who organised equipment, including underground tanks and pumps for a new petrol station, a food technologist who had helped Watties sell their baby food, and an electrical engineer working in the telecommunications industry). (Jennings, 1991, pp. 3–5)

We are aware of the myriad of factors that will influence girls' career choices, and that an SOS conference is just one of them. Our objective is to expose the girls to a high-quality experience. Throughout the two days that the conference runs, the participants are exposed to the following factors that contribute to positive feelings towards science.

Introducing a Range of Role Models

Research in Queensland by Eileen Byrne indicates that a critical mass of prominent women staff in scientific subjects is required before girls feel free to aspire to their roles (Byrne, 1993). We, too, acknowledge that we must be able to introduce a range of role models. Girls are not influenced by the lone inspiring woman scientist! Participation in science careers must appear to be the *norm* for their gender rather than the exception.

We encourage role models to talk about their families; often their young children will accompany them on the course. It is important to model the hidden agenda and show that it is possible to marry, have a family and be oneself while having a career of one's choice. We found that an interview at the end of the session by an organiser, or another student, promoted lots of personal questions such as: 'What sort of clothes do you wear on the job?', 'Do you feel lonely working on site?', and 'What was your favourite subject at school?' This personalised the role models for our form 4 students.

Creating a Friendly Non-Threatening Atmosphere in Which Young Women Are Considered Special

The conference atmosphere is fostered by providing, for refreshments, special areas that are furbished with flowers and comfortable seating. The venue is selected for its non-school atmosphere. We also give out folders containing the course information, name tags, badges and so on. Although there is an element of competition, we give *lots* of prizes. We also try to ensure that creativity and good communication are rewarded, as well as the acknowledged 'best' solution to the problem.

Demonstrating That There Are Alternative Images of Science

We aim to dispel the fragmented, objective and value-free image of science normally promoted within the curriculum (Brickhouse *et al.*, 1990). Kelly (1985a) comments that the 'masculine' image of science could be altered if 'packaged' in a way that appeals to girls. However, she qualifies her stance by noting that the link between masculinity and science is complex and involves the identification of a pupil's gender identity with her/his perception of the school subject. She argues that not only science but also scientific practice and science education are perceived as masculine.

Although I have sympathy with this view, I believe that SOS can contribute to an image of science in the working world in which women have a place, and can show that scientists work best when they cooperate and communicate effectively. In promoting SOS participation in schools, we have used 'advertising' such as:

When you think of science you might think of the three Ms – Mathematics, Mechanics and Macho. After SOS, I hope that you will link science to the three Cs – Communication, Creativity and Cooperation.

More holistic views of science (Bell, 1988) emphasise the ways in which science and technology can influence people's lives. Medicine and environmental technology are two of these fields. I like to quote the example of Justine Johnstone Wanger, who was part of the team that developed the slow-intravenous drip method of administering drugs, and Pearl Kendrick and Grace Eldering's invention of the whooping cough vaccine and the triple antigen vaccine, which protects most infants in the developed world (Stanley, 1983).

Modelling the Importance of Cooperation and Shared Learning Experiences during Problem-Solving Exercises

The emphasis of SOS is on cooperation rather than competition. Prizes reward originality as well as the most efficient and effective solutions. The presentation of group solutions in a whanau (family) atmosphere mean that feelings of inadequacy are minimised and that cooperation, as well as communication skills, are valued. Some of the activities involve a reporting session. For example, *The Choicest Jelly* activity includes time for the groups to report their marketing plan for this new improved dessert. Less-than-delicious samples are presented with great originality: 'Buy Jello the slimmer's jelly – it wobbles but you won't!'

Providing Extra Time for Girls to Experience Practical Skill Development

Because SOS concentrates on doing rather than listening, a variety of techniques are used to help girls work on their technological expertise. For example, each problem-solving exercise contains a variety of activities and tasks, mastery of a specific skill is rewarded with a proficiency certificate, learned helplessness is counteracted by allowing plenty of time for tinkering and for developing manipulative skills, and careful attention is paid to the distribution of equipment. In the *Eggstraordinary* exercise (which involves making an eggbeater from LEGO), it is important to distribute the LEGO pieces in separate bags, so that girls feel that they can attempt a solution on their own. If the LEGO is distributed in a communal bag, a dominant member often will commandeer the lot!

Illustrating That Science Is Relevant to Young Women's Lives

On the hidden agenda are the ideas that a working knowledge of science is important for all members of our society, especially when they are expected to make decisions that will affect their lives (Bell, 1988). These decisions can relate to, for example, the siting of a rubbish dump, alternative methods of sewage dispersal or the use of potentially-damaging chemicals by industry. The role models are encouraged to discuss the value judgements that they are expected to make within their job. Environmental engineers need to be aware of the community's needs, as well as the environmental and conservational issues when developing a recreational area. Mechanical engineers cannot look always to the cheapest and most efficient solution. In each case, the role models emphasise that an informed community influences the decisions made by business. In fact, businesses make unilateral decisions at their peril!

Although many of our students will not be continuing with tertiary science education, we hope to establish that a scientific education will give them a stake in informed decision-making within the community. It is our belief that an important goal of science education is responsible citizenship.

THE LOGISTICS OF PLANNING AND RUNNING AN SOS CONFERENCE

What was it like to organise the first SOS conference in Wellington? 'Fantastic – a nightmare' was the Head of Department's response (Jennings, 1991). This type of conference involves much forward planning and is impossible without the support of the school community. We have outlined the stages in the development of our workpack and videotape. However, some ideas are worthy of amplification, if only to illustrate the importance which we give to the complete package.

Sponsorship

Because this type of format requires money, sponsorship occupied much of our time until we had run our first SOS conference and produced the videotape and workpack. Since then, the programs have been self-supporting, with schools contributing to the running costs and the sale of workpacks and videotapes providing funds for meeting administrative costs. Our workpack includes an outline of the reasons for adopting this package. When a school has decided to run the program, we spend time discussing sponsorship development as well as strategies for involving the whole staff and the local community. We have found that the long-term planning is crucial in order to allow time for sponsorship and community participation.

Developing the Conference Format

School children and teachers are used to doing things on the cheap! Whenever we introduce the conference idea to a school, there is always the problem of finance, and teachers immediately think of ways to cut costs. We believe that money spent on so-called superficial trappings means that the atmosphere and special nature of the program is developed. Can one feel special in a drafty, shabby school hall?

Attention is given to the venue, preferably away or isolated from the normal school activity. Flowers, comfortable seating areas and refreshments can help to foster the right atmosphere. Each girl is issued with a folder which contains copies of the exercises, name tags, an SOS badge, a printed conference program and any stationery or 'freebies' that we have managed to glean from our sponsors.

The pace of the conference is fast, with each activity taking between an hour and an hour and a half. At most, there are four activities each day. We plan and organise the materials to the last sheet of paper, so that the change-over between activities is fast and the girls only have time to concentrate on the current problem. After all, their time is too important to waste.

Training of the Role Models

One must remember that the role models for these kinds of programs are not trained teachers and, therefore, it is very important to show them what is expected. Usually they come armed with a sheaf of notes, but discard them with alacrity when we explain their role. We have developed a training session which explains our objectives, shows a training videotape, runs through all the exercises, familiarises the team with the logistics of the program, and emphasises that they are the facilitators and that the teachers are the 'gofers'. It is important that the role models understand the importance of their role in the conference and that we encourage them to form their own group networks during the conference.

Course tutors are given time within the program to relate the exercises to their own careers, as well as to meet the girls informally. SOS conferences have invited a wide range of role models, including civil engineers, food technologists, computer engineers, architects, foresters, fitters and turners, systems planners, and mechanical and environmental engineers.

Empowering Teachers

In developing the two-day course, we have utilised many girl-friendly science strategies (e.g., role play, problem-solving, peer tutoring) and have emphasised the importance of teamwork and communication. In addition we have tried to demonstrate that science, and especially science in industry, values imagination and creativity.

We hope that, by the end of the conference, there is a group of teachers with some more ideas about how to introduce girl-friendly strategies into their own science classrooms, and that they will be willing and able to advise other schools. Although SOS was designed as a package, we are aware that it will be developed and used in a variety of ways. Our main objective is to empower educators to use this intervention strategy when and how they see fit. In fact, this approach has resulted in SOS conferences being held thoughout New Zealand with networks of teachers developing their own formats.

Developing Job-Linked Problem-Solving Activities

We do not claim that the activities are all original, but we hope that they have a particular 'female bias'. Not only are the problem-solving activities given a 'female-interest' contextual setting, but also the importance of talking and planning is emphasised. Thus, the teams that communicate are more likely to win a prize.

The activities are graded from quite simple construction exercises that provide hilarity and positive reinforcement, to those that require more technological and manipulative skills. By the end of the conference, the girls are tackling these problems with ease. Within the problem-solving design, we emphasise that there is always more than one solution to a problem. Our problems are located, like our role models, in food technology, mechanics, electronics, environmental engineering, architecture, and design and marketing.

WHERE TO FROM HERE?

Is the energy, time and commitment that is necessary for running a successful SOS conference warranted? Is this a good investment of our time? At times, our team can wonder, especially in the midst of the frantic activity of an SOS conference. But the informal responses from our students, both during and long after the conference, make plain that this intervention program was a highlight of Auckland Girls' Grammar School form 4 (13–14 years) science education.

We went back to Auckland Girls' Grammar to interview our first SOS clients when they were in form 7. These were some typical responses:

It was fun! It didn't feel like a chore and related more to life than what happens at school. I learnt heaps about cooperating with the group. I remember setting up the company and especially enjoyed the electronics. (Kate)

It was extremely memorable and was more fun than anything else because it wasn't boring and you were learning a lot. The jelly was the most fun. (Clare)

We had a crazy group and we had lots of fun. Our 'Ooh Ah Jelly' was a mess. Kate's drawing of the nail was not to scale and made an object which was four times larger than it should have been! (Jemma)

No matter how complimentary these comments might be, we realise that intervention programs need to be accompanied by complementary teaching strategies and activities. Like Alison Kelly, we believe that teachers' *active* cooperation is critical (Kelly, 1985b). Not only must an intervention program be part of an overall strategy of promoting girl-friendly science within the school curriculum, but also it must include a program of teacher education. Girls leave SOS inspired with a new vision of science, yet face the reality of the classroom where they are marginalised and where the traditional male view of the scientific world is presented. Even those committed and aware teachers also are limited by the ogre of the publicly-examined School Certificate at the end of form 5 (14–15 years).

In relation to teachers' needs, we realised that there was a dearth of New Zealand-based contextual material in Science with an SOS philosophy. The SOS team and Karen Mitchell (a secondary science teacher) wrote, developed and trialled a program which covers force, mass and acceleration (Mitchell *et al.*, 1992). The program, 'Motorway Madness', was set in the context of motorway accidents. A workpack and videotape describe the stages in development and delivery of a teaching package with filmed sequences of the teaching strategies being implemented. The teaching program concludes with students developing a publicity campaign on motorway safety. The class as a whole then discusses the campaigns, in the context of 'Motorway Madness', with a woman traffic officer participating in the discussion.

The SOS team continues to be committed to disseminating the workpack and videotape (Farmer *et al.*, 1990), developing new problem-solving exercises and a range of follow-up suggestions for teachers. Liz Godfrey has taken over the role as consultant facilitator in the Auckland area and has developed a New Zealand-wide database of women involved in science-based jobs as well as a support network of teachers with experience in running SOS courses. She also has liaised with the Electricity Corporation of New Zealand (ECNZ) to produce a series of eight posters which show women working in industry as well as aspects of their family life and hobbies.

Many schools question the need for a two-day SOS conference. A variety of courses have been attempted, ranging from one day to two hours! We realise that schools are constrained by small budgets and unsympathetic staff, and that there is the temptation to try anything rather than ignore the problem. To organise an SOS conference makes formidable demands on a busy teacher. However, we feel that the principal objectives of the course (to make *girls* feel special in science and to feel that they have a future in the subject if they so wish) can be achieved only if one has the space to develop cooperation and self-worth. And this takes time!

Like many other organisers of intervention programs, we have realised that merely to show girls the delights of a scientific career is insufficient. Many other obstacles, including peer pressure, parental aspirations and an unfavourable economic climate (that sees equal pay and opportunities too high a price for women's inclusion), have to be overcome if women are to pursue career opportunities in the physical sciences. Perhaps SOS is just one way of broadening young women's horizons.

Auckland College of Education, Auckland, New Zealand

REFERENCES

Bell, B.F. (1988). 'Girls and science', in S. Middleton (ed.), *Women, education and schooling in Aotearoa*, Wellington, New Zealand, Allen and Unwin, 153–160.

Brickhouse, N.W., Carter, C.S. & Scantlebury, K.C. (1990). 'Women and chemistry: Shifting the equilibrium towards success', *Journal Chemical Education* (67), 116–118.

Byrne, E. (1993). *Women and science: The snark syndrome*, London, Falmer Press.

Farmer, B., Godfrey, L., & McCowan, L. (1990). *SOS – skills and opportunities in science for girls, workpack and video*, Auckland, New Zealand, Auckland College of Education.

Geddes, R. (1991). 'Crisis – what crisis? Public awareness and the science-technology crisis', *New Electronics*, May, 8–9.

Jennings, M. (1991). 'Skills and opportunities for girls (SOS): Wellington's first conference', *WISE Wellington Newsletter*, June, 1981.

Kelly, A. (1985a). 'The construction of masculine science', *British Journal of Sociology of Education* (6), 133–154.

Kelly, A. (1985b). 'Changing schools and changing society: Some reflections on the girls into science and technology project', in M. Arnot (ed.), *Race and gender: Equal opportunities policies in education*, Oxford, Open University, Pergamon Press, 137–146.

Ministry of Education (1991). *Education statistics of New Zealand*, Wellington, Research and Statistics Division.

Ministry of Education (1993). *The status of girls and women in New Zealand education and training*, Wellington, Learning Media.

Mitchell, K., Farmer, B., McCowan, L. & Godfrey, L. (1992). *Motorway madness, workpack and video*, Auckland, New Zealand, Auckland College of Education.

O'Neill, B. (1990). 'Who wants to be an engineer?', *New Scientist* 126(1715), 24–29.

Rivers, M-J., Lynch, J. & Irving, J. (1989). *Project FAST, report of a national pilot study of future aspirations, subjects and training of senior secondary school students*, Wellington, NZ Department of Education.

Stanley, A. (1983). 'Women hold up two-thirds of the sky: Notes for a revised history of technology', in J. Rothshild (ed.), *Machina ex dea: Feminist perspectives on technology*, The Athene Series, Oxford, Pergamon Press, 5–21.

Sturrock, F. (1993). *The status of girls and women in New Zealand education and training*, Wellington, Learning Media.

NANCY KREINBERG[1] AND SUE LEWIS[2]

15. THE POLITICS AND PRACTICE OF EQUITY:
EXPERIENCES FROM BOTH SIDES OF THE PACIFIC

This three-part chapter provides a personal account of our experiences with gender equity initiatives in mathematics and science education in the USA and Australia. We have shared the satisfaction of being involved in the creation, development and ongoing success of programs in our respective countries. We were fortunate to spend a year working with each other in 1990, comparing and contrasting our programs and cultures. This has given us a fresh look at our own work and insights for future efforts.

In Part I of the chapter, we discuss our specific programs within the context of the broader issues of equity in our societies. In Parts II and III we provide more detailed analyses of the EQUALS project (in California, USA) and the McClintock Collective initiative (in Victoria, Australia). Each of these two parts gives some specific illustrations of many of the issues raised in the present chapter. We feel that the manner in which these issues are resolved will play a role in shaping the future directions of the two programs and, perhaps also, of schools as a whole.

PART I: THE LIMITS AND POSSIBILITIES OF EQUITY PROGRAMS

Over the last decade, awareness of gender issues in science and mathematics has increased significantly in both the USA and Australia, equity programs have become credible, and gender concerns are included routinely in educational policy documents. Equity practitioners have learned much about the layers of complexity that comprise gender issues in our societies and classrooms.

We have observed significant differences in the Australian and USA educational systems, but similar conditions occur for females in mathematics and science education. Both USA and Australian girls and women remain disadvantaged and underrepresented in mathematics and science courses and employment at all levels. In this context, one of our programs (EQUALS in California) began in the mid-1970s and the other (the McClintock Collective in Victoria) began in the early 1980s, celebrating, in its name, Barbara McClintock's 1983 Nobel prize for medicine and physiology. The original mission of both programs was to improve girls' and young women's participation and success in mathematics (EQUALS) and science (McClintock) courses. Both programs began because the majority of girls and young women were excluded from mathematics and science, and the prevailing stereotypes were hostile to their full participation. Staff in the programs struggled to achieve awareness of gender issues in their respective environments and to create advocates among teachers, working initially with a network of teachers to provide an additional source of expertise.

177

L.H. Parker et al. (eds.) Gender, Science and Mathematics, 177–202.
© 1996 *Kluwer Academic Publishers. Printed in the Netherlands.*

EQUALS and the McClintock Collective have experienced considerable success, although on different scales, by a number of measures: acceptance by teachers, administrators, and the state's educational policymakers; impact on curriculum and inservice education; awards of public funding; and increase in public awareness of the issues of women in mathematics and science.

In both mathematics and science classes, we have seen a change in the climate of teaching and learning in Victoria and California, which in our view is due in part to a changed gender perspective. For example, EQUALS and the McClintock Collective were early advocates and practitioners of cooperative learning. Research and classroom work in both Australia and the USA (Cohen, 1986; Dalton, 1985; Fennema & Meyer, 1989; Lockheed & Klein, 1985) indicated that girls were more comfortable and successful in classrooms where cooperation rather than competition was practised. Further, gender analyses illuminated classroom dynamics, including the importance of wait time, using questions, valuing students' contributions and background experiences, and equal sharing of resources in the classroom (Fennema & Peterson, 1987; Good & Brophy, 1987; Kelly, 1981; Sadker & Sadker, 1982). All of these understandings became part of our programs. In addition, on the basis of advice from practitioners, we advocated specific teaching strategies. These included use of problem-solving, hands-on materials, investigations, writing, role playing, discussions and collaborative groups. They have helped to make the learning of mathematics and science more accessible and relevant to all students, and particularly to females.

We concluded also that the reasons for the low participation of girls and women in science and mathematics lay not only in the teaching and learning of mathematics and science, but in the curricula as well. Understandings of gender issues clarified the nature of the hidden curriculum: mathematics and science education reflected the perspectives, values, and background experiences of the dominant culture. In mathematics texts, there are few problems or examples that reflect female experiences or non-white cultures. In science teaching, assumptions about students' backgrounds are based on boys' experiences with mechanical tinkering. The curricula themselves are not gender-neutral.

In science, the programs moved towards context-driven rather than content-driven curriculum. The McClintock Collective recast the curriculum so as to start from the context of girls' and boys' lives and move towards the building of scientific ideas through activities (Gianello, 1988). This is a reversal of established science curricula that start with theories and end with applications in industry or the environment. For example, traditional secondary units on light would start with theories about light and equations. The McClintock Collective's curriculum begins with students' questions and dilemmas about sunburn and coloured T-shirts, and builds towards scientific understandings through the students' experiences (Harding, 1986; Lewis & Davies, 1988).

Gender, Class and Ethnicity:
Questions for EQUALS and the McClintock Collective

We know that gender issues do not exist in isolation. We feel that it is imperative that equity programs combine in some way the insights of class and ethnicity. An understanding by teachers of the interplay of each is necessary to enable more children to be successful learners.

While the numbers of white, middle-class females in mathematics and science courses have increased in the past decade, a similar change has not occurred for low-income and ethnic minority students in both countries. One must question whether issues of gender, class and ethnicity ever can be separated.

When EQUALS and the McClintock Collective began, they questioned educational practice from a gender perspective. Early in its existence, however, the EQUALS program expanded its focus to include ethnicity issues to accommodate two main needs: teachers who were working with minority students and the challenges that ethnically diverse classrooms presented to teachers; and, funding priorities which, during the 1980s, moved rapidly from a focus on women to a focus on minorities.

EQUALS has struggled in its programs to integrate issues of class and ethnicity with those of gender. Many participating teachers arrive at an EQUALS workshop anxious about their own mathematics learning as well as angry about the social injustices in their own and their students' lives. To participate in discussions and activities concerning both requires courage and risk-taking on the part of staff and participants. It also requires a staff and participant group that reflects a diversity of class and ethnicity so that meaningful, rather than superficial, discussions can take place. A program can be strengthened greatly when complex issues of gender, class and ethnicity are combined, but there can be dangers of trivialising issues, obscuring messages or losing valuable insights that come from the individual perspectives of gender, class and ethnicity. Continued efforts are needed to serve equally the interests of gender, class and ethnicity.

In contrast to the Californian context, the McClintock Collective in Victoria evolved in an atmosphere in which gender, class and culture perspectives were well articulated in the education climate (Connell *et al.*, 1982). What was missing was the translation of those ideas into science curriculum and action strategies for classroom teachers. The McClintock Collective made a conscious decision to start with a gender focus because of the personal experiences of its members. Externally, it was not pushed as much as the EQUALS project by forces of diversity in education. Funding was provided solely on the bases of its gender profile.

Can and should the McClintock Collective continue to maintain its single-issue perspective, and what are the gains and losses in such a decision? Single-focus programs are often powerful in the clarity with which issues can be presented and understood. However, we know that people who construct curricula and programs reflect their own values and background experiences in this construction. This consequence could provide the impetus for the McClintock Collective to link and work with people from other ethnic and socioeconomic groups. Without this linkage, the Collective inadvertently might create a sexually-inclusive science curriculum to the exclusion of class and ethnic perspectives.

Limitations of Current Equity Practice and Consequences of the Hidden Curriculum

Equity programs have had a positive influence on a small number of teachers at the classroom level, helping them to change their practice and be alert to sexist and racist texts and practices. However, most classrooms remain teacher-dominated and content-driven rather than context-driven, with a white, middle-class orientation. Equity programs have not affected the hidden curriculum at a level beyond individual classrooms. We have not seen the kinds of changes, on a large scale, that ensure equity across schools, districts and states. We have not seen a positive change in the gap between affluence and disadvantage that could be the result of educational reforms. In fact, we have seen the gap increase dramatically in the USA and remain static in Australia. We have not seen political decisions in education that lead to a serious rethinking of established educational practice at the system level.

The hidden curriculum also operates at the system level. Here it functions to maintain the status quo that is at odds with equity practice. Equity seeks change to create more access for students outside the dominant culture (Parish *et al.*, 1989). The lesson for both the McClintock Collective and EQUALS has been to understand the power that the hidden curriculum wields and the need to look beyond the classroom and the gender issue to the broader context of schooling. In taking this broader look, we must determine what contributions we can make towards broadening equity: how does gender relate to equity issues shaped by class and ethnicity?

The Urban Schools Challenge: Focus on the System

The ways in which gender, class and ethnicity issues are interwoven are most evident in urban schools. Both the McClintock Collective and EQUALS work closely with urban teachers. As urban classrooms in each country become more diverse, fewer students are experiencing success, and fewer parents are pleased with the education that their children are receiving.

There have been attempts in both countries to repair this situation by concentrating on 'fixing' the students and the teachers. In the USA, massive amounts of money have been put into Chapter One programs, designed to provide remedial education for 'disadvantaged' (low income and minority) students. There have been created a few good programs that benefitted some students for a period of time, but the achievement gap between affluent and low-income students widens each year, as attrition and failure rates increase.

In Australia, similar initiatives occurred with the Disadvantaged Schools Program and the Participation and Equity Program, which were federally-funded attempts to develop alternative teaching and curriculum remedies. These programs experienced considerable local success despite short-term funding, limited resources and the vagaries of political changes. Yet, no long-term changes occurred in the way in which schools were organised.

Many now conclude that the systems themselves are at fault, and not the students or the teachers. Yet, the tendency is still to blame people, as if 'parents were sending the wrong kind of children to school' (Hart, 1989, p. 239) or teachers were not as dedicated as they once were.

The failure of inner-city schools is not an anomaly; it is a warning of the failure of an entire system. Albert Shanker (1990), president of the American Federation of Teachers union, wrote that 'we've reached the limits of our traditional model of education [and] . . . we can expect neither greater efficiency nor more equity from our education system'. Nonetheless, change will be difficult because it's 'easier to think in terms of improving the kind of school we all know than it is to imagine totally different kinds of schools' (p. 345). In this chapter, we share some of the insights into this problem which our work with EQUALS and the McClintock Collective has given us.

Teachers face equity issues every day. If students come from diverse ethnic or racial backgrounds, they are helped to understand and respect languages, values and experiences that are different from their own. If students are from the same racial background, there are issues of gender bias and stereotyping, ability grouping and socioeconomic status. If teachers have navigated these challenges successfully, their students still might be treated inequitably by other children, teachers, parents or administrators. These experiences also affect their mathematics learning.

More than any other school subject, mathematics serves as a race, sex and class filter. There is power and status associated with the study of mathematics. For those who achieve in the subject, there are substantial rewards in the opportunities available for entrance into well-paying fields of study and work. Any absence of equity in the teaching and learning of mathematics has severe repercussions on individuals and society.

A 'Critical Filter' Identified

In the early 1970s, the USA feminist movement was in its prime but the issue of women and mathematics did not surface until Lucy Sells (1975), a graduate student at the University of California, Berkeley, became curious about the relationship of secondary mathematics preparation to females' subsequent choice of a major in the University. She surveyed a small number of students on the Berkeley campus and found that women students had far less mathematics preparation during their high school years than male students. She was the first to call mathematics the 'critical filter'. She found that women were unable to select from a majority of majors at the University that required calculus and/or statistics, and that women were not being prepared to enter a vast range of mathematics-based fields or work. Moreover, because of their lack of quantitative training, women were far more vulnerable to being locked into low-paying, dead-end jobs after graduation. Students who were not college-bound equally were disadvantaged by inadequate mathematics preparation, because most of the apprentice programs in the skilled trades had mathematics entry tests.

The stereotype that girls were neither good in mathematics nor interested in science and technology was so pervasive in the culture that many reasonable people never questioned it. Very few educators had taken notice of the impact that lack of mathematics preparation had on girls' and womens' subsequent choices of work and study, and even fewer felt a need to try to change the 'natural' course of girls' interests towards language and the arts.

Sells began a personal campaign to convince people that the attrition of females from mathematics was a phenomenon that began early in girls' lives and had major impacts on their life choices. She pointed out the small numbers of women engineers and scientists in our country and challenged educators and the public to redress this inequity.

In the early 1970s, neither national nor state enrolment data in mathematics by gender had been gathered. California was one of the first states to do so, and it was found that females were underrepresented severely in advanced mathematics courses in high school. Later, in California, data were collected by ethnicity and it was found that minority females and males were even more disadvantaged by lack of participation in mathematics courses.

Concurrently, a major mathematics education reform was beginning in California, with an emphasis on problem-solving and a more interactive style of teaching. Many mathematics teachers who went on to become leaders in the state professional organisations became involved in the 1970s in programs that attempted to make mathematics more accessible to all students.

The EQUALS Project

Development of EQUALS: The Early Programs

EQUALS at the Lawrence Hall of Science, a public science centre at the University of California, Berkeley, was one of the first USA programs to inform teachers and parents about the importance of mathematics learning for girls and women and to provide methods and materials for teachers to use in their classrooms to promote female participation in mathematics. The development of EQUALS began in 1973, when we created a Math for Girls class, at the same time bringing to the public's attention the need to involve girls, as well as boys, in mathematics activities. An examination of enrolments in the Lawrence Hall's after-school classes revealed that less than 25% of the students were female; parents were sending their boys to the Hall for enrichment activities in mathematics and science, but not their girls.

Math for Girls, like a number of other classes at the Hall, introduced children to mathematical puzzles and games and attempted to develop their problem-solving abilities. For this class, however, only female teachers (Berkeley students who were majoring in mathematics or science) were used and they made a point of articulating the usefulness of the mathematics that the children were learning by relating it to a variety of occupations. The class became immediately popular and still is being offered as a way to introduce girls to the pleasures of mathematics.

As news of the Math for Girls class spread, people interested in the issues of women's participation in mathematics and science began to contact the staff. Soon, we were in the centre of a rapidly growing network. We invited local people who taught in colleges and universities, elementary and secondary schools, as well as scientists, engineers and parents, to meet with us and decide how to encourage more girls and women to become involved in mathematics and science throughout their schooling.

The result of these discussions was the formation of a volunteer organisation called the Math/Science Network. The first activity of the Network was to create a conference called Expanding Your Horizons to introduce young women (ages 12–18 years) to careers in the sciences and to women working in a range of scientific and technical fields.

The first conference, held in 1976 on a local college campus, was so successful that others soon were started throughout the country. Now, each year, approximately 70 such conferences are held in more than 30 states and in Australia, and they reach many thousands of students, parents and teachers with the message that women are definitely a part of mathematics and science.

These were satisfying efforts, but insufficient. It was clear to us that the only way to make a significant impact on the mathematics education of girls and young women was to work directly with teachers. Thus, EQUALS – a mathematics equity program to help K-12 teachers retain more female and minority students in mathematics – was born in 1977.

EQUALS Begins

The EQUALS course is 36 hours and is held in six sessions over an academic year. It includes instructional strategies to encourage cooperative learning, curriculum materials to promote problem solving, classroom activities that emphasise a hands-on approach to understanding mathematics concepts, career information to stimulate interest in mathematics and science, role models to motivate students to persist in mathematics and science courses, discussions of gender, class and culture issues, and models for instituting change in classrooms and/or schools to generate more equitable learning opportunities for all students.

Participants, approximately 75% of whom are female, come typically from inner-city, suburban and rural school districts serving K-12 students. In 1983, we began to establish sites in states outside California and today 38 such sites provide EQUALS programs in their communities. Each year, we offer a follow-up program called SEQUALS to maintain contact with EQUALS teachers and share our new ideas and activities with them.

Policy and Practice

The interrelationship of EQUALS practice and mathematics education policy exists on several levels. EQUALS is entrenched firmly in the mathematics organisation of the State of California. Staff members have served (and are serving) as officers of the California Mathematics Council, the statewide affiliate of the National Council of Teachers of Mathematics (NCTM), and are active in NCTM national committees. There is full support for EQUALS from the major professional mathematics organisations which means EQUALS presentations appear at most statewide and many national conferences, reaching thousands of new educators each year.

EQUALS is also anchored in the reform curriculum. The activities and instructional techniques that we provide reflect the philosophies of the California *Framework* of problem-solving, alternative assessment, cooperative learning and heterogeneous groupings.

Mathematical Power for All

While EQUALS was created to focus on gender in mathematics, similar problems were occurring with African-American, Latino and Native American students. The expectations held by teachers for non-white, non-Asian students in mathematics is low, students often see the subject as irrelevant, and they have experienced failure so often in the subject that they are convinced that they cannot do it.

While many similarities exist between gender and racial attrition from mathematics, other factors are different for non-white students. The schools which they attend often are primed for failure (Parish *et al.*, 1989). Social and economic circumstances usually are cited for the low-income, racial or ethnic minority students who fail in mathematics. However, 'we seldom say that a teacher is culturally deprived, that a school is at risk, or that a school district is disadvantaged or backward. The child and the family are labelled' (Cuban, 1989, p. 782). Rather than continuing to try to make the child fit the system, which we know is designed for white, middle-class students, we need to alter the system to respect and encourage children's differences while still preparing them to survive and succeed in a society that could be at odds with their cultural values.

Two of the assumptions currently under scrutiny at all levels of education in the USA are 'streaming' and 'assessment'. Both of these are basic to equity issues in education and are examined and analysed in EQUALS workshops.

Heterogeneous Groups

In the last few years, the notion of 'streaming' (placing students into different groups by perceived ability levels) has come under attack. Originally, it was intended to enable teachers to provide the most effective and efficient instruction. It also is based on the premise that certain students are destined for higher education and others are not. After decades of experience, it is clear that ability grouping, and the streaming to which it leads, dooms the majority of non-white students, as well as low-income whites, to low achievement and a lack of access to educational opportunities that drives them further into disadvantaged status (Oakes, 1985). Moreover, research has demonstrated that students in heterogeneous groups succeed as well as those in homogeneous groups, and that the net gain for both female and minority students is significant (Cohen, 1986).

It is not accidental that doubts about streaming in the USA have increased as our classrooms have become more diverse. Educational reformers now believe that it is necessary to unstream both mathematics classes and the mathematics curriculum so that equity of access and outcomes can be achieved. 'Changing demographics have raised the stakes for all Americans. Never before have we been forced to provide true equity in opportunity to learn' (National Research Council, 1989, p. 19).

Teachers in EQUALS discuss and think about the implications of unstreaming their classrooms with others who share their concern. For those who are committed to changing their curriculum and their instructional methods to create more equitable classrooms, there is no easy way out. It is difficult and scary, and they often will experience frustration and failure. There will be many who will tell them that it can't be done. But once they begin, they can see the potential for change.

Much has been written on ways to create heterogeneous groupings in all subject areas. Nationally, inservice programs exist to assist teachers who are willing to experiment with creating cooperative learning groups in heterogeneous classrooms. Because of the popularity of the subject, many school districts are offering their own programs. Most teachers who have begun to do this warn that it is a slow process, with inevitable setbacks, but the results are worth it.

How does one begin? First of all, not alone. In EQUALS workshops, people attend in teams from the same school so that support is built in. Naturally, it is best to have administrative and parental support, but often that comes later, when the sceptics are convinced that achieving equity will not compromise excellence but, in fact, promote it. EQUALS teachers might be amongst the few teachers at their school who are willing to begin to experiment with ways to make their classrooms more encouraging of the diversity that each child brings to school, whether that is race, language, gender, ethnicity or academic, physical and social skills.

Diversity has to do with the way in which children learn, as well as their race, class or culture. Traditional methods of instruction rely on one method of learning: memorisation of facts in particular sequences ordered in a rigid timeframe of school periods and semesters. Children, like adults, learn in many different ways and learn best under various time constraints. Part of what teachers value in EQUALS is the opportunity to experiment with different approaches and get feedback from other participants about effective strategies.

Alternative Assessment

'Historically, assessment has played the role of legitimizing the disabling of minority students' (Cummings, 1986, p. 29). More than any other country, the USA relies on standardised tests to determine what students are learning. These tests emphasise rote learning and computation and do not provide an accurate picture of students' understanding of mathematics concepts beyond computation. 'Overemphasis on improving test scores inevitably means that what is tested will be taught, and that what is not tested will not be taught' (Stenmark, 1989, p. 31). Furthermore, those students who do poorly on standardised tests become more fearful and negative towards mathematics. Our traditional testing methods serve only a few at the expense of many students.

As long as tests drive the curriculum, and teachers are pressured to teach to the tests, we will not have a mathematics curriculum that is rich and flexible enough to provide access for all students. An emphasis on heterogeneous grouping of students, with a problem-solving curriculum, requires an assessment of achievement that is as complex and varied as students themselves. Reformers now are experimenting with alternative ways of assessing student understanding, including open-ended test items, portfolios of students' writing, drawing and thinking in mathematics, observations, interviews, videotapes and written reports of students' work.

One EQUALS publication is a handbook on alternative assessment (Stenmark, 1989) that has been disseminated widely and is used as the major introduction within the EQUALS workshop to the topic of assessment. As with other EQUALS publications (Downie *et al.*, 1981; Erickson, 1986, 1989; Fraser, 1982; Kaseberg *et al.*, 1980; Kreinberg, 1977; Stenmark *et al.*, 1986), these materials provide a means to extend the EQUALS philosophy and methodology beyond the workshops to a wider public.

Outreach to the Community: A Family Response

EQUALS was not designed to address the needs of parents who wanted to help their children with mathematics, especially those parents who themselves were unprepared in the subject. EQUALS teachers began to ask us for help in getting parents involved in their children's mathematics education. Many of these parents were in low-income and/or minority communities and did not participate in school activities. Our response was to develop FAMILY MATH in 1981 to help parents and children to learn and enjoy mathematics together. Twelve years later, the program has classes occurring in 46 states of the USA, Australia, Canada, Costa Rica, New Zealand, Puerto Rico, South Africa, Sweden and Venezuela.

FAMILY MATH courses – taught by a parent or teacher in a school, community centre or church – give parents and children opportunities to develop problem-solving skills in mathematics and to build an understanding of mathematics concepts. The program has served to increase the access of low-income, minority and non-English-speaking families to mathematics. Weekly parent/child classes involve families in problem-solving activities, using hands-on materials, learning collaboratively, meeting role models, and discussing mathematics concepts and the importance of mathematics to future options. Weekly classes last for two hours on average, and go for from four to six weeks. They are led by volunteers (teachers and parents) who attend FAMILY MATH leadership workshops to prepare them to teach classes. Many national sites offer leadership workshops. The project staff at EQUALS direct the national dissemination and continued curriculum development.

In 1985, federal funds provided the means to establish sites in five community-based agencies serving Latino, African-American and Native American families in California, Oregon, Indiana, Arizona and Washington, DC. Publication of the FAMILY MATH book in 1986 and the Spanish translation in 1987 stimulated growth of the program. In 1989, Matemática Para La Familia was initiated specifically to serve Spanish-speaking families in California, Arizona and Florida.

Innovative features of the program include parents becoming teachers of mathematics and collaborating with teachers to present classes, teachers becoming school leaders through involvement in the program, using household materials to teach mathematics, providing activities that can be extended and repeated at home, and modelling a way for parents to help their children enjoy and understand mathematics.

A federally-funded evaluation of the impact (Shields & David, 1988), as well as local evaluations (Devaney, 1986; Kreinberg & Thompson, 1986), indicate a high degree of acceptance of FAMILY MATH methods and materials by families of diverse race, culture, ethnicity and socioeconomic status. Many teachers reported improved understanding of mathematics by students in FAMILY MATH classes, and parents indicated more positive attitudes towards their involvement in school activities as a result of the program.

We have learned that, when it is made easy and comfortable for parents to help their children with mathematics, even the most mathematics-anxious parents eagerly will become involved. We have found that the FAMILY MATH materials work with children ages 5–15 years and with adults from diverse backgrounds. We have seen attitudes of both parents and children change from dislike, to cautious acceptance and to enjoyment of mathematics over the span of a few weeks in a FAMILY MATH course. We also have seen parents who have not been involved in their children's schooling begin to participate because of FAMILY MATH.

Eighteen Years Later

In 1977, when we first began EQUALS, women comprised less than 1% of the engineering workforce; today they comprise 8%. For all the efforts in the last 18 years, the gains for women in scientific and technical fields have not been substantial and, yet, we do not feel that we have failed. Over the years, the EQUALS program has grown from a part-time effort on the part of four people to a permanent staff of 20, with satellite centres in 31 states, a publication series of 15 books, and several spin-off programs that have established reputations in their own right. Over 75,000 teachers in 36 states of the USA and several other countries have participated in EQUALS programs. A number of evaluation studies have documented the program's effectiveness with participating teachers and students (Kreinberg, 1989; Sutton & Fleming, 1987; Walsh & Hirabayashi, 1988; Weisbaum, 1990; White & Conwell, 1987).

Among the program's accomplishments has been a significant increase in the awareness of parents and teachers about the importance of mathematics and an increase in female course enrolments. People concerned with attrition of African-American, Latino and Native American students from mathematics find EQUALS methods and materials appropriate for assisting underrepresented students in mathematics. Most importantly, teachers have found the program to be a way of increasing their own competence and confidence in the teaching and learning of mathematics and enhancing their professional development.

Nevertheless, the effort that has been expended is on too small a scale to make the changes that are necessary. We know what needs to be done. We have progressed beyond the stage of defining and studying the problem to developing, testing and implementing strategies and programs to increase female and minority participation in mathematics and science. The results have been impressive, but the numbers of students who benefit from these programs remain disturbingly small. Program models that work have been transferred to other locations with promising results, yet the number of scientists and engineers who are female or minority is still alarmingly low. It is not a problem that can be solved within the school system as it now exists.

Future Directions

It is clear to most Americans that the public school system no longer works for a large number of students, most of whom are poor, of colour and do not speak English. This population will comprise the majority of our public school students in the next decade. Alternatives to our current system of education must be found quickly.

It is in this state of crisis that equity educators have the most to offer. Those of us who have been working in equity for years know that new approaches to old problems are possible, and that teachers are willing collaborators in new approaches as long as their knowledge and experience are honoured.

Teachers who are drawn to alternative methods of teaching and learning exist in every school at every grade. They are the ones who are challenged by a difficult situation, curious about a better way to teach, bored by a routine that no longer is useful or compelling. Teachers who embrace change are ones who know that every child, every day, presents a new situation to be understood. Nothing can work all the time with all students. Responsiveness to children's needs is the bottom line for these teachers. However, the school system, as currently established, does not support deviations from the norm.

What teachers are learning in EQUALS is that there is no one way to teach mathematics and no one way to learn it. Similarly, there is no one way to educate all children; teachers must be free to develop and learn best how to reach each child. They need the structure to do this — a system that honours innovation, teamwork, cooperation, experimentation and meaningful accountability.

Trusting the Teacher

We need to begin trusting teachers to do what is right and helping them to find out what that is by setting up the conditions in which they can discover what to do from each other, from their students, and with staff developers. We must trust them and, at the same time, not expect them to perform miracles under conditions that sabotage their efforts.

Equity practitioners can set the conditions under which this can happen by continuing to question the education reforms being promulgated and by providing a forum for teachers to do the same. Teachers can ask of each reform effort: How does it enable the system to become more responsive to diversity, more open to innovation, more enabling of teachers' decision-making power? In what ways is the reform perpetrating old myths and beliefs that have served to maintain the status quo?

As teachers question the system within which they are working together, they are more likely to see in what ways they can effect change, with whom they must work to do so, and what outcomes they require. Helping people to do this together, rather than individually, is essential equity practice. Substantive changes will not occur if limited to single teachers and individual classrooms. Teachers need to be able to see how what they do makes a part of the whole and that they have a potential, with others, to transform what no longer is useful into what is needed.

Equity practitioners embody a faith in and respect for teachers. Our programs are built upon a belief that teachers have strengths that have not been tapped, knowledge of their craft that can enable all students to be successful, and the commitment to do so. We assume that, given the necessary support, teachers can and will act as equity advocates on behalf of all students. What is clear, however, is that teachers' dedication and commitment alone cannot solve our educational problems. We must be willing to push for something that doesn't yet exist — a system of schooling that is truly equitable for all students.

The task facing us is enormous. We cannot expect teachers to create equitable classrooms alone, nor can we do it without them. The issues of equity are as important as any that we face today.

In contrast to Part II's focus on EQUALS in the USA, this part of the chapter describes the origins of the gender and science movement in Victoria, and reviews the inclusive strategies that have developed, particularly those associated with the McClintock Collective. Using an adaptation of the Schuster and van Dyne (1984) curriculum change model, the chapter evaluates some of the Victorian gender and science directions over the past 10 years. It is primarily a case study of the work in Victoria, but it includes references to major federal events.

If the analyses and feminist theories surrounding science education are to have any practical use, they must reach the classroom. It simply is not enough to get more women into science and wait for the 'critical mass' to transform science. If we continue to train women and men in science in the traditional ways, then we are likely to see the continuation of a science disconnected from social and environmental concerns. In this context, one of the achievements of the Collective's work is that, as an organisation, it has persisted through time, acting as a clearinghouse, network and focus for cumulative experiences and understandings in this complex curriculum area. Through the maintenance of the Collective, there is a nucleus for the storage of experiences as well as an organism for critically analysing the approaches over time.

Origins of the Gender and Science Education Movement in Victoria

Equal Opportunity Strategies: Access to Non-Traditional Jobs for Girls and Changing Classroom Dynamics

Inner urban schools are often the source of innovative educational ideas because they are the places where students from economically and socially stressed homes (where English is often a second language) are failed by the conventional curricula and assessment criteria. A radical education movement flourished in inner urban schools in Melbourne in the 1970s and continued into the 1980s with the establishment of an Equal Opportunity Project in schools in the Melbourne suburb of Brunswick (BRUSEC, 1982).

The Victorian Ministry of Education in 1977 established an Equal Opportunity Unit which provided vital support for the gender and science projects as they emerged. The Transition Education for Girls Project, funded by the federal government's Transition Education Advisory Committee, was located in the Equal Opportunity Unit in 1980 and already was focussing on programs for girls to enter non-traditional jobs, analysing classroom dynamics, and publishing materials aimed at putting 'missing women' into science and mathematics resources for schools.

The equal opportunity movement was at the edges of the mathematics and science subject areas. However, science and mathematics courses soon became a focus of the Brunswick Equal Opportunity project, because they attracted few girls at a time when teenage unemployment and school drop out rates were high. The project also conducted 'Try A Trade' days which gave girls an experience of a range of trade options in order to broaden their thinking when they planned for their future. In hindsight, the non-traditional jobs programs that were offered were based on trying to persuade girls to try the jobs that boys were doing. For some girls, however, the step of seeing that you could have a future job in any area was a prior hurdle to being able to value these experiences.

At that time, there was little analysis of the nature or future of the trades that we cheerfully were inviting girls to enter. However, girls who chose these jobs were not destined to have much fun on the shop floor, as the sexism and loneliness would be difficult to bear. The work values and exploitation that some of these workplaces reflected for both female and male employees were not considered seriously in these early equal opportunity programs.

The other major strategy of the equal opportunity movement in the early 1980s was the examination of classroom dynamics. The attention of teachers was directed to the hidden and not so hidden gender biases in their interactions with students. There was interest in who was asked a question, the duration of wait time between the teacher's question and the student's answer, the time teachers spent with which students, and the quality of time spent with whom (Kelly, 1981; Stanworth, 1983). The equal opportunity movement could translate classroom dynamics into action by monitoring classroom interactions between teachers and girls and boys. These analyses provided the bases for change strategies such as asking more girls for their input, ignoring the first hand up, spending more time with the girls, and monitoring for equitable distribution of resources.

The job and school situation also was stacked against many boys from working-class and non-English speaking backgrounds. While the gender analysis provided some action strategies for girls, class-based analysis of the plight of low-income students (girls and boys) understandably was struggling to provide a framework for action by teachers. It seemed that educational equity issues were much easier to define in relation to gender than in relation to class. Increasing awareness about classroom dynamics and the contribution of boys' social under-development assisted in forming ideas about the changes needed in science curriculum and teaching.

The Start of Collective Action in Victoria

In the context of these beginnings, the first statewide inservice program for teachers on gender and science was organised in November 1983 and, significantly, was titled Equal Opportunity in the Science Classroom. This focused on the inequalities for girls in science classroom practices, in curricula that were based more on the experiences of boys, and in textbooks that represented boys and men doing science and girls absent or passively watching. This first workshop examined the inequalities facing girls in their schooling and the employment disadvantages that followed. The inservice course provided a timely meeting place for educators who had been working in isolation and were concerned about the participation of girls in school science programs. On that day, the McClintock Collective was formed and, ever since, there has been a monthly meeting devoted to organising, supporting, working, understanding, discussing, politicking or writing about gender and science. The membership network grew quickly to over a 100 science educators, and it has been a connecting thread through all the gender and science education programs in Victoria.

Many of the early McClintock Collective members had been the women enrolled in science courses in the early 1970s, who had watched the initial impact of feminism from a university cafe, while their language arts sisters were immersed in the debates and discussions. Others were feminists who were trying to integrate their science-trained selves with their feminism. Still others approached the issue from a social justice viewpoint as well as from an educational belief that the traditional teaching of science was inadequate for many girls and boys. An important role served by the Collective has been the development of teaching strategies which are inclusive for girls and boys. A culmination of this earlier work is the Collective's *Getting into Gear — Gender Inclusive Teaching Strategies in Science* (Gianello, 1988), which was published in 1988.

The McClintock Collective provides professional development activities and collaborative curriculum development projects for teachers. It has been represented on most of the major science education committees, policy bodies and curriculum development projects in the state. In 1991, plans for a Girls and Maths and Science and Technology Centre in Melbourne came to fruition with the establishment of Hypatia's Place, the first such initiative in Australia.

The Collective's work has been achieved largely by members whilst employed in their science education jobs. More recently, there has been effective collaboration with a state-funded project officer position. Most members are secondary and tertiary teachers rather than primary teachers. State funding for inservice work has provided time release for members to conduct and facilitate inservice programs in collaboration with the project officer. One early grant was for the training of the Collective members in workshop facilitation skills; this was important because workshops often contained participants who were antagonistic and there was a need for approaches that were not alienating to these participants.

An important feature of the development of the McClintock Collective was that it was based on the principles of collaboration and non-hierarchical structures. Initially, there were no papers to be published, no fame to be found, no research money available, no egos to be flattered and no high-powered high-salaried jobs to be filled. Instead, there were many years of networking and time-consuming discussions about the same issues from newly developing perspectives. Many hours were spent contacting people and encouraging their involvement. Writing submission after submission for funding for inservice workshops is a hidden but time-consuming task. Telephoning politicians and bureaucrats and fighting yet again to secure minimal funding has not furthered many Ministry careers. These are some of the aspects of the Collective work that need to be recognised: important collaborative work that often has been regarded as 'women's work' has formed the very successful foundation and ongoing strength of the Collective in a world with few collaborative models.

Policy Documents in Gender and Education

Since the start of the Collective's work, a range of policy documents important in assisting the gender and science education programs have been developed. After an era of general reports and documents on the situation for girls in Australian schools (Australian Schools Commission, 1975; Commonwealth Schools Commission, 1984), the then Commonwealth Schools Commission (1987) took up the challenge of developing The National Policy for the Education of Girls in Australian Schools. At the same time, the Commonwealth Schools Commission and the Australian Science Teachers Association funded the development of the Girls and Women in Science Education Policy (ASTA, 1987). Both policy documents involved the national and state education communities in an extensive consultative process and were as important in the discussion processes which they generated as they were in their content.

The initial excitement surrounding the development of these policy statements has not been matched by action in the post-publication stages. There are many sound and imaginative strategies for science curriculum and teaching reform that have not been taken beyond the pages of the two policies. Neither document can enforce accountability requirements and there has been minimal funding to develop the initiatives further. However, they do exist and practitioners have built on these commonly-expressed beliefs and exposed the funding gap between political pronouncements and practical implementations.

Towards a Gender-Free Science Curriculum and Teaching: Are We Challenging the Paradigm Yet?

Models of curriculum change appearing in the feminist literature (McIntosh, 1984; Schuster & van Dyne, 1984) provide an important reference point for the analysis of gender and science programs in Victoria and elsewhere (Lewis, 1993). All these change models depict the incorporation of women into the curriculum as a process with identifiable phases. McIntosh (1984) and Schuster and van Dyne (1984) developed two similar models from analyses of arts curriculum changes. Both models start from the historical situation in which the absence of women in the curriculum was not noticed, through to the transformed inclusive curriculum in which women's and men's experience can be understood together. Schuster and van Dyne (1984) also include the teacher relationships with the student in the course that each phase requires. They document a process describing how many teachers and students experience the process as curricular change through a gender analysis. I have applied these two frameworks to the processes of the Victorian movement, although the sequence and terminology are weighted more towards the Schuster and van Dyne stages. In the following discussion, the stages are (1) the absence of women not being noticed, (2) the search for the missing women, (3) why there are so few women in science, (4) studying woman's experience in science, (5) challenging the paradigm of what science is, and (6) the transformed, reconstructed gender-free curriculum.

Stage 1: Absence of Women Not Being Noticed

Because of the equal opportunity and non-sexist movement of the 1970s and early 1980s, and an increasing number of Australian women scientists, the first stage of 'not missing the women' had passed in science education by the time the McClintock Collective started in 1983. There are few women in the physical sciences in particular in Australia and this pattern is evident in schools and society at large. The numbers of women working in professional science positions peaked during the second World War and subsequently has been lower. However, even in the 1980s, science education in Victoria still has remnants of this 1960s stage, in which the Einsteins are the only great scientists, the students are the 'vessels' and maintaining 'standards' is the evangelical motivation. Knowledge was assumed to exist free of social and political biases.

Despite broad social recognition that women are missing from science and that social justice requires action to change this, the Collective still finds that there are many objections to these data being aired: 'Why are you forcing these girls to do something in which they are not interested?', 'I do not even want to think about changing the way in which I have taught physics for 20 years!', 'Girls simply are not interested in physics so save your energy!', 'I do not want to face the possibility that what I have been teaching might have stopped girls pursuing physics'. These responses reflect difficult issues that we must continue to address. In general, the only teachers who are exposed to the gender and science dilemmas are the teachers who are concerned about their girls; these teachers commonly are referred to as the 'converted'. Diminishing funds for teacher professional development have restricted the number of these opportunities for teachers in recent years. There is an ongoing need for more bridges to be built with 'non-converted' teachers who currently are not concerned with these equity issues.

Stage 2: The Search for the Missing Women

The search for the missing women scientists to profile in science classrooms was a phase that started in the early 1980s in Victoria with the publication of the resource package Women in Mathematics and Science Kit (Transition Education Girls Project, 1982). As Schuster and van Dyne (1984) point out, this approach adds to the existing data within the conventional paradigms. Essentially it was a compensatory exercise involving identification of the great women scientists. It transferred the same criteria that defined men as famous scientists, forgetting, for example, the women who did all the technical work in research laboratories who rarely were given credit.

Other authors have reframed and redefined the 'great women' in question either by documenting women who were not recognised, such as Rosalind Franklin (Sayre, 1975) and, at least until 1983, Barbara McClintock (Keller, 1983), or by recognising the 'nameless' women who work in industrial and research laboratories everywhere. Many Australian projects have published poster series, identikit broadsheets and curriculum activities featuring 'non-famous' women in science and teaching occupations.

Stage 3: Why Are There So Few Women in Science?

Questions such as 'why are there so few women in science?' framed much of the McClintock Collective's early work and exemplified the first 1983 inservice program for teachers in Victoria. The absence of women in science employment, and of girls in physics and chemistry classes, were profiled. These gender-based differences were attributed to social and educational influences.

Some of the responsibility was assigned to the particular classroom dynamics experienced by girls in the science laboratory classroom. All too often, boys grabbed the equipment, demanded the teacher's time and used their previous experience in informal science to answer the questions before the girls. Textbooks used examples that were more familiar to the informal science experience of boys and sexist language that excluded girls. Whilst these sexist classroom dynamics and resources are important, they now are recognised as being the first layers of the gender and science onion. The inner layers, consisting of the curriculum itself, teaching practice and the view of science which they both reflected, were only starting to be challenged in Victoria in the mid-1980s.

We were asking why there were so few girls in science classes and, whilst we had some convictions that science and not girls had to change, we often interpreted these data within the existing paradigm of science. Women were disadvantaged because individual females missed out on being part of male achievement. It was a protest rather than a direct challenge to the conventional paradigm of science and science education.

Stages 4 and 5: Studying Women's Experience in Science and Challenging the Paradigm of What Science Is

In Schuster and van Dyne's (1984) curriculum change model, there are separate stages of studying women on their own terms and challenging the conventional paradigm in a discipline. In the Victorian gender and science work, these phases are intertwined both in memory and in the records. For instance, there was a huge shift in directions for the Collective in late 1984 when 30 members took off to the coast for four days to write a new gender-inclusive curriculum in science. Not unexpectedly, in hindsight, we spent most of the time attempting to clarify what a gender-inclusive science curriculum might be.

What is women's experience of science and what are the informal science experiences of girls? What are girls' views of scientific phenomena? What are the differences amongst girls? How does an inclusive curriculum differ from the traditional curriculum? Do girls learn differently from boys and what are the implications for science education? What are curriculum materials for girls? How do you start from female experience in science but extend girls from there? What is the influence of how these materials are taught? What areas of the science curriculum are most urgently in need of better materials? What are the implications for female and male science teachers? These and other questions took hours of debate and were discussed more over the next three years (McClintock Collective, 1987).

Out of this experience, our major focus became how science is taught, and numerous versions of a list of McClintock teaching strategies evolved. Many of these were present in other science education projects but often without a gender perspective. Teaching strategies included cooperative group work, creative drawing and writing, role playing, media, social implications of science, discussion, tinkering, negotiation, using materials and tools, values clarification, visits and visitors, negotiating the curriculum and sex role awareness raising (Gianello, 1988; McClintock Collective 1987). Essentially these teaching approaches fitted one of three types: they provided active learning contexts for students (e.g., constructing with LEGO); they described alternative ways of organising the classroom (e.g., cooperative groups); or they reorganised the curriculum (e.g., starting from and valuing students' experiences).

The Collective's overarching categories reflected a different science in the classroom, involving communication, creative science, developing practical skills, personal growth and science as a human activity. There was increasing emphasis on the science contexts of girls' lives and on the ways in which girls prefer to learn. There was a strong intuitive base of our own preferred learning styles and a motto that 'science should be fun'. Negotiating with students, valuing students' thinking and valuing the diversity of students' experience were all concepts that spread out from these explorations. After three years and many hours of discussion, these inclusive curriculum discussions and inservice programs were published in *Getting into Gear* (Gianello, 1988). This book traced the rationale for the work of the Collective through to 1988 and its activities emphasised both content and process, that is, the type of science that is taught as well as the way in which science is taught.

Perhaps the work of the Collective that gets closest to challenging the dominant science education paradigm, and which is one of the inner layers of the gender and science onion, is the fundamental construction of science curriculum that has come out of the work of Jan Harding (1986) and the Collective. Harding (1986) worked with students in teacher training to develop a curriculum that begins with students raising questions about the science issues in their lives. This issues-driven curriculum starts from, for example, lead in petrol, household appliances, acid rain, supermarkets or roller skates. Through investigating and checking with other people and ideas, this curriculum moves to a construction of scientific ideas within the starting framework of the students' previous knowledge and experience. This issues-driven approach is very different from the textbook-driven and content-driven science courses that still are operating in the majority of science classrooms. The lack of connection between science and the lives of students often is a repeated criticism of science by girls (Kearney & MacDonald, 1987; Department of Education and Science, 1980).

Embedded in this context approach to science education is the radical view that the curriculum needs to be negotiated with the students and that teachers have to hand over some control for what directions the curriculum takes.

Constructing curriculum in this way can challenge the dominant paradigm in science through the inclusion of all the issues that surround students' lives, from in vitro fertilisation to nuclear power. Science classrooms can be places where the practices of science are discussed, different views expressed, alternative information considered and local, national or international action initiated.

With the Schools Commission funding of the Girls and Maths and Science Teaching (GAMAST) project in 1986, there was a teacher inservice program with the specific aims of supporting primary and secondary science and mathematics teachers, assessing the school context in relation to the participation of girls and boys, developing teaching strategies for science and mathematics classes that provide success for all students, and documenting their learnings for other schools (Lewis & Davies, 1988). The inservice model of the GAMAST project was based on teams of interested teachers working together for a year with the support of consultants, asking their own questions, collecting data and designing a program for teaching an inclusive mathematics or science curriculum. Many of the inclusive strategies of active learning, different classroom organisation and the reorganisation of the curriculum were developed further by the teachers in GAMAST. The GAMAST project reinforced the value of action research for teachers as well as the importance of empowering teachers to design and implement change collaboratively. The other clear message was the reminder of how slow change is within the vicissitudes of school environments.

Stage 6: The Transformed, Reconstructed Gender-Free Curriculum

The gender-inclusive curriculum movement in science education focussed on the inclusion of female experiences and perspectives to change curricula and teaching. These analyses restrict the discussion of gender perspectives to what is constructed socially as female or male. The analyses do not include the broader political and social critiques of science and society, nor do they envision a world where gender is not an organising category. There is a need for a transformed curriculum that goes beyond the restrictions of female and male, that examines critically the assumptions behind the culture and practice of science and, as pointed out by Evans in Chapter 6, the social construction of masculinity and femininity.

Where Is the McClintock Collective?

In terms of the Schuster and van Dyne (1984) model, the McClintock Collective is hovering between Stage 5 and the beginning of Stage 6. Many questions about our vision of science need new initiatives and theoretical frameworks (Bearlin, 1987). We need to develop ways of articulating critiques of dominant scientific ideology and develop ways of incorporating these perspectives in the classroom. There is a steady increase in girls' participation in some areas of science, but science itself often is unchallenged.

From the feminist critiques of science we must strive for a curriculum which uses all that we have learned so far about humanising the classroom and curriculum, and providing students with skills to challenge a science that often is environmentally destructive and disconnected from human needs. Unless we do this, I think that we will not be furthering the important journey of reclaiming the practice of science and science education from within.

Many questions need to be explored in the inner layers of the gender and science onion in pursuing these curriculum changes. How can we value the diversity of female and male cognitive styles? How can we work constructively, when our power and professional status are invested in traditional science structures? How can we include other classes and cultures that science also has excluded? What sort of science and science practices do we want to promote in developing countries? How do we make sure that the practices, ethical questions and future dilemmas of science are explored with students? These are the questions to which we now seek answers, to guide us towards the implementation of a gender and culturally inclusive science education in the 21st century.

Conclusion: Rethinking the System

One of our major conclusions from our experiences, on both sides of the Pacific, with equity programs is that we need to rethink most of our cherished assumptions of schooling: the one teacher/one-grade 'eggcrate' style of teaching; the non-teaching administrators; the need for continuous testing of students; the length of the school day and year; the sequencing of topics in the curriculum; and the use of textbooks. None can be held sacred and all must be open to examination if a real change in education is to occur, rather than a tinkering or patching up of what is inadequate.

Equity activists can play a crucial role in this rethinking. We can demonstrate models of teaching and learning that *do* work with students outside the dominant culture and help teachers to understand and value all students. We can assist schools to operate in very different ways as equity is put into practice. Simultaneously, we must press for changes in the system that will enable such models to thrive. Currently, equity models are only marginally successful because they are at odds with the traditional system.

At first, equity practitioners illuminated biases in the curriculum and teaching of mathematics and science and identified changes that had to be made. As teachers went through EQUALS and the McClintock Collective programs, they were encouraged to think about how the approach to learning which they were experiencing (activity-based, cooperative, problem-solving) affected their students. They experienced the process themselves, reflected on its meaning for their own classrooms, and considered the action steps needed for implementation. Now, they must be asked to think about the political implications of this process. It is in this next step that both programs need to develop more understanding, vision and expertise.

We must assist teachers in transferring equity understandings at the classroom level to the school organisation levels. Our challenge is to create workshops which, while focused on mathematics and science education, also enable teachers to gain political insights and strategies for creating alternative ways of schooling. This is new territory for our mathematics and science equity programs. As we continue in our work, we must find new ways of communicating the difficult messages about equity; these messages are difficult because they challenge established beliefs, are complex to understand and require innovative workshop experiences.

Equity education is political education. Equity practitioners envisage schools where students and teachers work in small, democratic communities, spending several years together, sharing decision-making about curriculum and instruction. If schools were to be rethought and recreated along these lines, the knowledge and experience gained from equity practice would flourish. Schools then would serve and respect students and teachers from many backgrounds, cultures, classes and experiences.

ACKNOWLEDGEMENTS

The writing of this chapter would not have been possible without the contribution of the other members of our respective programs. Margaret Bearlin, Victoria Foster and Colin Hocking also are to be thanked for their interesting discussions that contributed to the ideas in this chapter.

[1]University of California at Berkeley, USA;
[2]Swinburne University of Technology, Melbourne, Australia

NOTES

1. Editors' Note: Nancy Kreinberg, founding director of the EQUALS program, has provided this analysis.
2. Editors' Note: Sue Lewis, a foundation member of the McClintock Collective, has provided this analysis.

REFERENCES

Australian Schools Commission (1975). *Girls, school and society*, Canberra, Australian Government Printing Service.
Australian Science Teachers Association (ASTA) (1987). *Girls and women in science education policy*, Sydney, Author.
Bearlin, M.L. (1987). 'Feminist critiques of science: Implications for teacher education', in J.Z. Daniels and J.B. Kahle (eds.), *Contributions to the Fourth International Gender and Science and Technology Conference, Vol. II*, Ann Arbor, MI, University of Michigan, 145–152.
Brunswick Secondary Education Committee (BRUSEC) (1982). *Equal opportunities project 1982 report*, Melbourne, Transition Education Advisory Committee (TEAC).
Cohen, E.C. (1986). *Designing groupwork: Strategies for the heterogeneous classroom*, New York, Teachers College Press.
Commonwealth Schools Commission (1984). *Girls and tomorrow: The challenge for schools*, Canberra, Australian Government Printing Service.
Commonwealth Schools Commission (1987). *The national policy for the education of girls in Australian schools*, Canberra, Australian Government Printing Service.
Connell, R.W., Ashenden, D.J., Kessler, S. & Dowsett, G.W. (1982). *Making the difference: Schools, families and social division*, Sydney, George Allen and Unwin.
Cuban, L. (1989). 'The "at-tick" label and the problem of urban reform', *Phi Delta Kappan* (70), 780–801.
Cummings, J. (1986). 'Empowering minority students', *Harvard Educational Review* (56), 18–36.
Dalton, J. (1985). *Adventures in thinking, creative thinking and co-operative talk in small groups*, Melbourne, Nelson.
Department of Education and Science (1980). *Girls and science* (HMI Series: Matters for Discussion 13), London, Her Majesty's Stationery Office.
Devaney, D. (1986). *Interviews with nine teachers* (Report for the FAMILY MATH Project), Berkeley, CA, Lawrence Hall of Science, University of California.
Downie, D., Slesnick, T. & Stenmark, J.S. (1981). *Math for girls and other problem solvers*, Berkeley, CA, Lawrence Hall of Science, University of California, .
Erickson, T.E. (1986). *Off and running*, Berkeley, CA, Lawrence Hall of Science, University of California.
Erickson, T.E. (1989). *Get it together*, Berkeley, CA, Lawrence Hall of Science, University of California.
Fennema, E. & Meyer, M.R. (1989). 'Gender, equity and mathematics', in W.G. Secada (ed.), *Equity in education*, New York, Falmer Press, 146–157.
Fennema, E. & Peterson, P.L. (1987). 'Effective teaching for girls and boys: The same or different?', in D.C. Berliner and B.V. Rosenshine (eds.), *Talks to teachers*, New York, Random House, 111–125.
Fraser, S. (1982). *SPACES: Solving problems of access to careers in engineering and science*, Palo Alto, CA, Dale Seymour Publications.
Gianello, L. (ed.) (1988). *Getting into gear: Gender inclusive teaching strategies in science*, Canberra, Curriculum Development Centre.
Good, T.L. & Brophy, J.E. (1987). *Looking in classrooms*, New York, Harper and Row.
Harding, J. (1986). 'A foundation chemistry course from issues', *McClintock Memos 5*, Melbourne, Hawthorn Professional Development Centre, 1–9.
Hart, L.A. (1989). 'The horse is dead', *Phi Delta Kappan* (71), 237–242.
Kaseberg, A., Kreinberg, N. & Downie, D. (1980). *Use EQUALS to promote the participation of women in mathematics*, Berkeley, CA, Lawrence Hall of Science, University of California.
Kearney, D. & MacDonald, J. (1987). *Don't step on my dream: Middle school physics*, Melbourne, Victorian Curriculum Advisory Board.
Keller, E.F. (1983). *A feeling for the organism: The life and work of Barbara McClintock*, New York, W.H. Freeman and Co.

Kelly, A. (ed.) (1981). *The missing half: Girls and science education*, Manchester, Manchester University Press.

Kreinberg, N. (1977). *I'm madly in love with electricity and other comments about their work by women in science and engineering*, Berkeley, CA, Lawrence Hall of Science, University of California.

Kreinberg, N. (1989). *EQUALS on-site: Annual project report* (Report to California Postsecondary Education Commission), Sacramento, CA, California State Department of Education.

Kreinberg, N. & Thompson, V. (1986). *FAMILY MATH: Report of activities September 1983 – September 1986* (Report to the Fund for the Improvement of Postsecondary Education, US Department of Education), Berkeley, CA, Lawrence Hall of Science, University of California.

Lewis, S. & Davies, A. (1988). *Gender equity in mathematics and science: The girls and maths and science teaching project*, Canberra, Curriculum Development Centre.

Lewis, S. (1993). 'Lessons to learn', in F. Kelly (ed.), *On the edge of discovery*, Melbourne, Text Publishing Co., 257–280.

Lockheed, M.E. & Klein, S.S. (1985). 'Sex equity in classroom organisation and culture', in S.S. Klein (ed.), *Handbook for achieving sex equity through education*, Baltimore, MD, Johns Hopkins Press, 189–217.

Malcom, S.M. (1984). *Equity and excellence: Compatible goals*, Washington, DC, American Association for the Advancement of Science.

McClintock Collective (1987). *The fascinating sky: Introducing the McClintock collective and some of its work*, Melbourne, Participation and Equity Program, Government Printer.

McIntosh, P. (1984). 'The study of women: Processes of personal and curricular revision', *The Forum for Liberal Education* 6(5), 2–4.

National Research Council (NRC) (1989). *Everybody counts: A report to the nation on the condition of mathematics education*, Washington, DC, National Academy Press.

Oakes, J. (1985). *Keeping track – how schools structure inequality*, New Haven, CN, Yale University Press.

Parish, R., Eubanks, E., Aquila, F.D. & Walker, S. (1989). 'Knock at any school', *Phi Delta Kappan* (70), 386–394.

Sadker, M.P. & Sadker, D.M. (1982). *Sex equity handbook for schools*, New York, Longman.

Sayre, A. (1975). *Rosalind Franklin and DNA*, New York, W.W. Norton.

Schuster, M. & van Dyne, S. (1984). 'Placing women in the liberal arts: Stages of curriculum transformation', *Harvard Educational Review* (54), 413–428.

Sells, L. (1975). 'Sex, ethics, and field differences in doctoral outcomes', Unpublished Doctoral Dissertation, University of California at Berkeley, CA.

Shanker, A. (1990). 'A proposal for using incentives to restructure our public schools', *Phi Delta Kappan* (71), 345–357.

Shields, P.M. & David, J.L. (1988). *The implementation of FAMILY MATH in five community agencies* (Report to the EQUALS Program), Berkeley, CA, Lawrence Hall of Science, University of California.

Stanworth, M. (1983). *Gender and schooling*, London, Hutchinson.

Stenmark, J.K. (1989). *Assessment alternatives: An overview of assessment techniques for the future*, Berkeley, CA, EQUALS and The California Mathematics Council.

Stenmark, J., Thompson, V. & Cossey, R. (1986). *FAMILY MATH*, Berkeley, CA, Lawrence Hall of Science, University of California.

Sutton, R.E. & Fleming, E.S. (1987). *EQUALS at Cleveland State University: 1985–86 evaluation report* (Report to the College of Education), Cleveland, OH, Cleveland State University.

Transition Education Girls Project (TEGP) (1982). *Women in maths and science kit*, Melbourne, Equal Opportunity Unit, Education Department of Victoria.

Walsh, M.F. & Hirabayashi, J.B. (1988). *EQUALS on-site (1977–1988): Evaluation of the replication efforts at five regional centers* (Report to EQUALS), Berkeley, CA, Lawrence Hall of Science, University of California.

Weisbaum, K.S. (1990). *Families in FAMILY MATH research project* (Final Report), Berkeley, CA, Lawrence Hall of Science, University of California.

White, R. & Conwell, C. (1987). *The impact of project EQUALS inservice: Testimony of participating teachers* (Report to Director), Charlotte, NC, Mathematics and Science Education Center, University of North Carolina.

LÉONIE J. RENNIE[1], LESLEY H. PARKER[1] AND JANE BUTLER KAHLE[2]

16. INFORMING TEACHING AND RESEARCH IN SCIENCE EDUCATION THROUGH GENDER EQUITY INITIATIVES

The purpose of this chapter is twofold. First, we present a cross-national comparison of the implementation of a gender equity initiative in Australia and the USA. Second, we provide an example of the ways in which both teaching and research in science education can be informed through the development and implementation of gender equity initiatives. The chapter documents the stages in a collaborative and evolving process involving two almost identical studies, one in Australia and one in the USA, and uses the combined results of the two studies in developing an explanatory model of the relationship between gender and science in schools and classrooms. Both studies used inservice workshops not only to address primary school teachers' lack of background knowledge and skill in teaching physical science, but also to give them training in gender-equitable teaching strategies. Both interventions were monitored and evaluated in terms of student and teacher attitudes, beliefs and behaviours in relation to science.

In the first part of the chapter, we describe the study conducted in Australia in 1983; in the second part, we document the study carried out in the USA in 1990–1991 and compare and contrast the findings of the two studies. In the third part, we describe the explanatory model derived from the combination and extrapolation of the findings (Kahle et al., 1993) and we illustrate a number of applications of the model with reference to earlier chapters in this book.

PART I: AN AUSTRALIAN STUDY USING A PERSUASIVE COMMUNICATION APPROACH

This study, like many of the others reported in this book, was conducted in the context of the reported under-participation and under-achievement of girls, relative to boys, in the physical sciences, and the subsequent disadvantage at which this places girls in relation to the pursuit of occupations and careers in a technological society (Kelly, 1978; 1981; Vockell & Lobonc, 1981). By the early 1980s, there was increasing awareness that the problem needed to be addressed early in children's education, before they were faced with choices that interacted with their development as adolescents and young adults. These views were supported by evidence that students' attitudes towards science and technology were formed during the primary school years (e.g., Ormerod & Duckworth, 1975). Clearly, those years were emerging as important ones during which to ensure that children experienced science in a balanced and equitable way.

Anecdotal evidence from primary school teachers, however, revealed that very

L.H. Parker et al. (eds.) Gender, Science and Mathematics, 203–221.
© 1996 Kluwer Academic Publishers. Printed in the Netherlands.

few approached science teaching with any confidence. Many confessed to feelings of inadequacy, anxiety and even dislike in relation to science and some tried to avoid teaching it altogether. Others were of the view that science was very much a boys' or men's area. There clearly was cause for concern about the likely effect of these kinds of attitudes, perceptions and expectations regarding young children's science education. Such concerns were supported by research which indicated that teacher's own beliefs and attitudes are a critical influence on students' attitudes towards science (McMillan & May, 1979; Simpson, 1978).

In collaboration with the state Education Department Superintendent of Equal Opportunity, Lesley Parker obtained a grant from the Australian Commonwealth Schools Commission in 1983 to develop, implement and evaluate an inservice program designed to facilitate a gender-equitable approach to the teaching of science in primary schools. The project, entitled *The Effect of In-service Training on Teacher Attitudes and Primary School Science Classroom Climates*, focused on teachers and students in fifth grade (aged 9–10 years) in the physical science topic of *Electricity*. The specific aims of the project were (1) to raise teachers' awareness of the adverse long-term effect on girls resulting from the general community's regard for science as an almost exclusively male domain, (2) to assist teachers to acquire skills in creating and maintaining an equitable science classroom environment and also to update their content knowledge and pedagogical skills in relation to Electricity and (3) to monitor the effectiveness of the program by assessing the nature and extent of attitude change and skills acquisition among teachers, observing the patterns of interaction within the classroom during the teaching of the topic, and assessing the nature and extent of any change in students' attitudes and perceptions about electricity.

Method

As described in detail in the project report (Rennie *et al.*, 1985), a teacher inservice program was designed to address the issue of equitable education in general, and the participation of girls in mathematics and science in particular. Participating teachers were selected on the basis of a questionnaire designed specifically for the project and administered to all fifth grade teachers in public schools in a mixed socio-economic region of the metropolitan area of Perth, Western Australia's capital city. Twenty teachers (five matched pairs of males and five of females), who had relatively low self-perceptions of their own skills and confidence in teaching physical science topics, were invited to participate in the study. All teachers accepted the invitation and one teacher from each pair was assigned randomly to a 'skills-equity' or a 'skills-only' group for inservice education.

The overall approach of the inservice program bore some similarity to the model of Shrigley (1983), based on the idea of persuasive communication and including the three concepts of 'persuade', 'mandate' and 'reward'. The 'skills-equity' inservice program was designed to occupy two separate days, of which one half day was devoted to skills for teaching Electricity. The teachers were led through a sequence of six activities, then given sufficient equipment to take back to their schools to implement the topic with their classes. In the other one and a half days, a variety of 'credible communicators' led and facilitated audiovisual presentations and discussions to raise teachers' awareness of sex-role stereotyping in the community, school and curriculum resources. For the one-week break between the inservice days, teachers were provided with worksheets and activities which required them to identify and examine gender-related issues in their own schools and classrooms. A range of techniques for overcoming tendencies towards sex bias were discussed with them, particularly in relation to differential expectations of boys' and girls' confidence and competence. The 'skills-only' group of teachers was provided with only the half-day workshop relating to the teaching of Electricity. These teachers, like those in the 'skills-equity' group, also were given sufficient equipment to carry out the activities with their classes back at school.

Following the inservice workshops, the teachers taught a sequence of six lessons on Electricity following the syllabus, approach and procedures presented at the inservice course. The approach that we advocated was 'hands-on' and inquiry-based, requiring students to work through the activities in small groups; currently these strategies are recognised more widely as being gender-equitable (see, for example, Chapter 13 of this volume). One male and one female teacher from the skills-only group elected not to teach this topic and took no further part in the project. Extensive data were obtained from the remaining 18 classes to assist in the monitoring of the outcomes of the inservice program. All classes were visited and observed. In addition, information was gathered by means of a second specially-designed questionnaire administered to the teachers after the completion of the Electricity topic, and from initial and final questionnaires designed for and administered to all the students.

Results

The full results of the study are available in the project report (Rennie *et al.*, 1985) and, because of their similarity to the findings of the study in the USA, details of comparison and contrast are documented in more detail in the second part of this chapter. An overview of findings from the teachers' and children's questionnaires, and from classroom observations, is presented here, followed by interpretations and judgements about the success of the project.

Teachers' Questionnaires

The initial teachers' questionnaire completed prior to the inservice workshops, as expected, revealed that relatively little science was taught at fifth grade level in primary schools (no more than 8% of total time). Female teachers, in particular, tended to have low perceptions of their own background knowledge and skills, especially in the physical science area. Teachers' general impression was that, in physical science topics, girls experienced more difficulty than boys, and girls' interest and confidence were less than those of boys. Also, teachers thought that girls showed less interest, experienced more difficulties, performed less well and had lower self-confidence in physical science than in biological science. Importantly, male teachers rated themselves more highly than female teachers in terms of their own interest, background knowledge and skill in teaching science, and teachers rated boys and girls in ways which were consistent with these self-ratings.

The final teachers' questionnaire, completed after the intervention, showed change in teachers' personal attitudes and beliefs. Compared to their responses to the initial questionnaire, all teachers were more positive about their interest, background knowledge and skill in teaching the Electricity topic. Teachers from both the skills-equity and skills-only groups perceived the Electricity topic to be useful and relevant to both boys and girls, and rated students as very interested and enthusiastic about the topic. Most male teachers perceived it to be equally easy for boys and girls, but some female teachers thought it harder for girls than for boys, and they perceived boys to have more confidence and a higher performance than girls.

Teachers were unanimous about the usefulness of the inservice workshops. They appreciated the opportunities to work through the learning experiences associated with the Electricity topic and the help given to them with lesson programming and equipment. They also reported that the inservice course had given them more knowledge and the confidence to try a 'hands-on' approach. Nearly all of the teachers in the skills-equity group considered that the inservice program had made a difference to the way in which they treated boys and girls. Most saw themselves as more conscious of girls' presence and girls' science-related needs, and as manifesting this increased awareness in more equitable discussions and in less patronising treatment of girls.

Classroom Observation

The lessons observed in the 18 classrooms participating in the study were successful in terms of the students' engagement 'on task' for an average 95% of lesson time, and in terms of their participation in group 'hands-on' activity work for about two-thirds of the lesson and in whole-class instruction for the remaining one-third of the lesson. In the whole-class situation, girls received more teacher-initiated interactions and boys initiated more interactions with the teacher. Questioning strategies were equitable in the skills-equity classes, where understanding and knowledge level questions were distributed equitably but, in the skills-only classes, boys received disproportionately more questions at the higher cognitive level. When the class was involved in group work, participation by the skills-equity group of teachers in the inservice course was associated with active involvement of girls, particularly when work groups were mixed-sex. Teachers in the skills-equity group supervised the mixed-sex groups very closely, perhaps accounting to some extent for the higher levels of active involvement by girls in their classes. Teachers tended to initiate teacher-student interactions with all-boys groups, while students tended to be the initiators of interaction with all-girl groups. Many girls appeared to lack the confidence to go about their tasks without reassurance from the teachers. This is consistent with the lower levels of self-confidence reported by girls on the children's questionnaire, and matches the lower levels of self-confidence in girls perceived by teachers in their responses to the first questionnaire.

Children's Questionnaire

The children's responses to the initial questionnaire revealed that, contrary to conventional wisdom, the girls and boys had similar average levels of interest in science. However, the patterns of their preferred activities differed, and it was evident that boys' and girls' levels of interest in various topics or activities appeared to reflect quite closely their likely previous out-of-school experiences. Boys had a greater preference for activities relating to electricity than girls did, and expected to enjoy the topic more.

The children's responses to the follow-up questionnaire showed that all classes enjoyed the Electricity topic, and a comparison with identical items on the initial questionnaire showed that the differences between boys and girls were much reduced. For the classes of the skills-equity group of teachers, sex differences averaged zero, but in the skills-only group the differences remained statistically significant, although they were smaller. The children reported that they enjoyed their work in Electricity and found it generally easy. In response to questions about their competence in handling the electrical equipment, boys thought that 'most boys' were good at it, and that they themselves were nearly as good as 'most boys', but they thought that girls were less able. In contrast, girls perceived 'most girls' to be as good as 'most boys' at working with the equipment, but they saw themselves personally as being less able than 'most girls'. Fewer boys than girls in each group thought that women could become electricians, and more boys than girls thought they could become electricians themselves. However, more girls in the classes taught by the teachers of the skills-equity group thought that they could be electricians than girls in the classes of teachers without equity training.

These results appeared to indicate that boys see themselves and other boys as being more capable than girls in the traditionally male field of electricity. Girls generally see other girls and women as capable in this field but see themselves as being less capable. Overall, the differences between boys and girls on nearly all items on the follow-up questionnaire were smaller in the skills-equity group than in the skills-only group. It seemed that, on average, teachers in the skills-equity groups were more able than teachers in the skills-only group to teach the Electricity topic in an equitable manner.

Discussion

The feedback from teachers, the observation of classroom interactions and the results of the children's questionnaires indicated that the aims of the project had been achieved, at least in the short term. It is possible to suggest some reasons for the success of the inservice workshops, particularly the equity aspects. Given that the project addressed systematically the crucial area of teacher attitudes and, as indicated earlier, that the approach adopted was similar to Shrigley's (1983) model of persuasive communication, this discussion is presented in terms of Shrigley's three components of 'persuade', 'mandate' and 'reward'. Certain elements of the 'persuade' component appear to have been critical to the success of the program as a whole.

First, as hinted earlier, the inservice component of the program employed a variety of highly credible personnel – an Education Department primary science advisory teacher, a Superintendent of Equal Opportunity, two qualified science teachers and a selection of researchers in the areas of attitudes to science, subject choice, classroom interaction and girls in science. This seems likely to have been important, given Shrigley's finding that expert and trustworthy sources are more effective than less credible sources in changing attitudes. Second, the 'persuade' component of the equity part of the inservice workshops contained a variety of activities which placed the issues in a broader context, through audiovisual aids developed elsewhere in Australia and overseas. Third, the immediate local relevance of the issues was emphasised, illustratively through the use of local research data on sex differences in subject choice and achievement, and practically through the session focussing on hands-on activities in an actual syllabus topic. Fourth, the 'persuade' component contained what emerged to be a critical time period, between the two inservice days, for structured reflection, self-analysis and observation. The reactions of teachers to this structured reflective period were very positive. They seemed, as a consequence, to find the remainder of the inservice program more personally relevant and most teachers became very committed to the program.

The program also contained incentives for these teachers to change their attitudes and classroom behaviours and, in this sense, provision was made for Shrigley's 'mandate' and 'reward' components. The mandate was provided through the inbuilt processes of evaluation throughout the project. All teachers knew from the outset that they were going to be observed and questioned about the extent to which they had been able to implement the strategies developed during the inservice program. While potentially this could have been perceived as coercive or threatening, it did not emerge as such, perhaps because of the non-threatening nature of the 'persuade' component and the attempt during the one-week reflection period to help teachers to feel a sense of ownership of the program.

Finally, with respect to the 'reward' component, there appeared to be a variety of ways in which the teachers gained pleasure and satisfaction from their participation in the program. One of these was simply the opportunity to discuss important issues with colleagues and experts away from the everyday pressure of the school and classroom. Other rewards came through their increased feelings of competence and confidence in relation to the skills and knowledge addressed in the program, and through the feedback about themselves and their students which was provided to them at regular intervals throughout the monitoring of the program.

Overall, the initial outcomes of the inservice program were encouraging. Teachers' attitudes appeared to have changed in the intended direction, with apparent positive effects on their behaviour and on their students' attitudes and behaviour. Based on this study, a preliminary schema representing the possible relationship between teacher attitudes, teacher behaviour, student attitudes and student behaviour was developed and presented at the third International Gender and Science and Technology Conference (GASAT 3) (Parker & Rennie, 1985). At the time, the intention was to test these relationships empirically, and perhaps present the results quantitatively, in a kind of path analysis. As events transpired (and as perspectives on research matured and changed), the initial study did not lead in this direction. It led instead to some exciting collaborative research and, ultimately, to the development of an explanatory model described in the following parts of this chapter.

PART II: A USA STUDY OF STRATEGIES FOR CHANGE

At GASAT 3, there was considerable discussion of the Australian study described in Part I of this chapter (Parker & Rennie, 1985). Subsequently, the opportunity arose for a considerable amount of collaborative work, expanding on the Australian study and involving researchers in both Australia and the USA. As part of that collaboration, Jane Kahle obtained funding from the National Science Foundation to carry out a second intervention study in the USA in 1990–1991 (Kahle *et al.*, 1991). This study, entitled *The Effect of Teacher Inservice Programs on Elementary Students' Achievement and Attitudes in Science*, followed the method of the Australian study closely. Its purpose was to examine the effectiveness of three types of teacher inservice workshops in challenging gender socialisation within the classroom. It focused in particular on students' attitudes towards science and the participation of women, and on the participation of boys and girls in physical science classroom activities. In order to facilitate cross-national comparisons, the study was designed to replicate the Australian one as much as practicable. However, adjustments to accommodate curricular and socio-cultural differences were made, and the research design was modified to accommodate more recent research findings, to provide a control group (skills only) and to analyse retention of any behavioural changes by teachers or students.

Method

The USA study was carried out in an urban suburb of a large metropolitan area in the Midwest. Like the Australian study, the sample of teachers and their students was drawn from schools serving families in a mixed socio-economic community. The design of the study was enhanced by having three, rather than two, inservice workshops, described as 'skills-equity', 'skills-only' and 'equity-only'. The two days of skills-equity teacher workshops matched those in the Australian project as closely as possible. A further design improvement was to include some additional aspects to supplement the equity-only and the skills-only workshops, so that each workshop lasted two days. Three differences between the studies were beyond the control of the Ohio researchers. Whereas the Western Australian study used matched pairs of male and female fifth grade teachers selected from a large school district, the Ohio study was constrained by smaller school districts, and so used all fourth and fifth grade teachers in one district. Because only five of the 23 teachers were male, it was not practicable to consider the sex of the teacher as a variable in the study. All teachers in a given school were assigned randomly to the skills-equity (eight teachers), skills-only (eight teachers) or equity-only (seven teachers) workshops. The compositions of the student samples in the two countries were different. All children in the Western Australian study were Caucasian, but about half (52%) of the Ohio sample were Caucasian, with African-American students comprising the other half.

The Ohio study proceeded in much the same way as the Western Australian study. Both teachers and students responded to initial and final questionnaires, which were the same as those used in Western Australia, except for minor wording changes to fit the American idiom. The workshops occurred after the completion of the initial questionnaires, followed by the teaching of the Electricity topic, using equipment and teaching notes provided during the workshops. As in the Western Australian study, the teaching approach advocated was hands-on and inquiry-based, and the same content was covered. Lessons during the teaching of the electricity topic were observed, using the classroom coding sheets developed during the first project. The final questionnaires for teachers and students were administered after the topic was completed.

Results

All three teacher workshops resulted in at least short-term changes in teachers' and students' attitudes towards the Electricity topic. Like those in the Western Australian study, Ohio teachers reported that little science was taught in the elementary school, and they rated their preference, skill and interest in teaching physical science lower than in other areas of science (i.e., biological and earth/space science). Ohio teachers also rated their teaching resources as less than adequate, and reported that their science lessons were predominantly teacher-directed rather than activity-based. Hence, the hands-on, inquiry orientation to the Electricity topic presented a change in the approach of some of the teachers, and observations revealed that students clearly enjoyed doing the activities.

Overall, in terms of the effects of the initiative, the findings of the two studies were strikingly similar. Details of particular aspects of the studies are provided in Bailey (1992), Kahle *et al.* (1991), Kahle and Damnjanovic (1994), Kahle and Rennie (1993), Parker and Rennie (1986), Rennie and Parker (1987) and Rennie *et al.* (1985).

Here, the results are summarised in several sections relating to teachers' and students' questionnaires and classroom observations.

Teachers' Questionnaires

Prior to the commencement of the workshops, both the Western Australian and the Ohio groups of teachers had similar perceptions of the usefulness, relevance and easiness of science for boys and girls, and of boys' and girls' performance in science. All features were rated highly. In both studies, biological science was perceived to be easier than physical science for both boys and girls. In Ohio, where earth science was taught, biological science was perceived to be easier than both earth and physical sciences. There also were some gender-related perceptions. For physical science, all teachers believed that boys had more confidence and greater interest than girls. For biological science, Ohio teachers rated boys more confident and interested than girls, whereas Western Australian teachers perceived girls and boys to be equally interested and confident.

Following completion of the teaching of the Electricity topic, when teachers were asked to reflect on boys' and girls' responses to the electricity lessons, the teachers in both countries rated both boys and girls highly with respect to the use and relevance of the topic and interest in the topic. Furthermore, they indicated that they perceived no differences between boys and girls in relation to those variables. However, irrespective of their treatment group, teachers in both studies rated boys as more confident than girls, but they noted that girls' confidence improved during the topic. Ohio teachers perceived boys and girls to have performed equally well in the topic and expressed surprise at girls' willingness and ability to do the hands-on activities. Western Australian teachers perceived boys to perform better than girls. In both studies, the only treatment-related difference was that regarding the easiness of the topic: teachers with skills-equity or equity-only training perceived boys and girls to find the topic equally easy, whereas teachers without equity training (the skills-only group) rated the topic easier for boys. Again, in both countries, teachers' ratings on items related to students' interest, confidence and performance in Electricity were much higher in the posttest than in the pretest.

Teachers in both studies judged the workshops favourably. In terms of their own background knowledge, interest and skills concerning electricity, the skills training was perceived to have enhanced these attributes. Teachers with equity training responded that they were more aware of girls' activities in the classroom, and about half noted that they made efforts to increase the girls' participation.

Classroom Observation

In both the Ohio and Western Australian studies, observation in classrooms focused on girls' and boys' access to resources and participation in the electrical activities, teacher feedback to students and teacher-student interaction. With respect to students' opportunities to access the curriculum materials, Ohio teachers were given enough equipment for students to work in pairs, so that all students had ready access to these resources. In Western Australia, however, there was sufficient equipment for groups of two to four students. In the latter case, access to resources was sex-differentiated by some teachers, allowing boys to dominate the equipment and (unintentionally) providing obstacles for girls who, initially, were less skilled in making circuitry.

In both Ohio and Western Australia, teachers gave little praise and very little criticism to students apart from indicating whether their responses were right or wrong. The distribution of teachers' feedback to students indicated no strong trends relating to the sex of the student and there were no variations associated with the type of inservice workshop which teachers attended.

Although the interactions between students and teachers were monitored in both studies, in Ohio there were insufficient interactions at a whole-class level to analyse. Observations of teacher-student interactions during group work revealed that, in both studies, teachers with equity training paid more attention to the working of mixed-sex groups and monitoring girls' and boys' participation in the activities. Teachers without equity training were more likely to use the equipment in all-girls' groups, demonstrating but also often completing the task for the students, thus depriving girls of the satisfaction of completing the activity. It was noted that the equity-only group of Ohio teachers, the third treatment group which had no skills training, interacted least with their students, perhaps suggesting a continued lack of confidence in their own science ability.

Students' Questionnaires

Responses to the student questionnaires in Ohio and Western Australia indicated that the extent and pattern of science interests for boys and girls were remarkably similar in both places, although the sex-related differences tended to be smaller in Ohio (Kahle & Rennie, 1993). In both studies, the traditional stereotypical pattern suggesting that boys prefer physical science and girls prefer biological science was not clearly evident in students' interests. Rather, students seemed to like those things with which they had the chance to become experienced. Overall, it was evident that girls expected to enjoy activities related to the Electricity topic much less than did boys.

Responses to the questionnaire completed after the topic were revealing. First, both boys and girls enjoyed the topic, with higher ratings from girls. Boys' ratings were similar regardless of the type of workshop that their teacher attended, but girls' ratings were noticeably higher in classes of teachers with equity training. Further, all significant sex differences favouring boys prior to the topic disappeared and, after the topic, the only differences favouring boys were in the skills-only classes. On the basis of the Western Australian results, Parker and Rennie (1986) pointed out that planning science instruction on the basis of students' expressed likes and dislikes could be inappropriate, because relative dislike of some topics could be associated with lack of previous experience. The similar, but more recent, results from the Ohio study underscore the importance of that advice.

In both countries, boys perceived the Electricity topic to be easier than did girls, regardless of which workshop their teacher attended. As noted earlier, in both countries, teachers with equity training considered that girls and boys found the topic equally easy. It also was noted that the Ohio equity-only teachers who received no skills training interacted least with their students during the lessons. It was in these classes that the average differences between boys and girls were greatest.

Students' perceptions of their own and others' confidence in doing the topic provided interesting contrasts. Boys in both studies perceived 'most boys' to be better than 'most girls' and themselves to be as good as 'most boys'. Western Australian girls perceived 'most girls' to be as good as 'most boys', but Ohio girls perceived 'most girls' to be better than 'most boys'. Girls in both studies perceived themselves to be not as good as 'most girls'. Clearly, boys have more confidence than girls in their own ability. The striking finding in the Ohio data was the large discrepancies between the positive self-rating of each sex's competence and the negative rating of the other sex's competence. Teachers' ratings of boys' and girls' performances, as well as classroom observations, suggested that these discrepancies were an exaggeration. Little, if any, difference was observed between girls and boys in their skills or performance.

The different composition of the Ohio sample allowed a sex by race analysis of the results for the students' questionnaires. Kahle and Damnjanovic (1994) found that both African-American and Caucasian girls enjoyed the activities, decreasing the sex differences evident on the first questionnaire. However, Caucasian girls had less self-confidence in their ability concerning the Electricity topic and found it to be more difficult than did the African-American girls or boys in both racial groups.

The career of 'electrician' is stereotypically male, and boys and girls were asked whether it was a career option for themselves and for women. Girls almost unanimously considered that women could be electricians, whereas about 80% of boys did. In Ohio, and in the skills-equity group in Western Australia, about 90% of both boys and girls thought that they could become electricians. However, only about 70% of girls, compared to 90% of boys, in the Western Australian skills-only group thought that they could be electricians. Stereotyping of this occupation is consistently greater in boys than girls.

Discussion

Despite the cultural and educational differences between Australia and the USA, and the seven-year time lapse between the studies, there were few differences between the findings of the two studies. Two differences relating to students' attitudes stand out: the smaller sex-differences in attitudes and interests on the intitial questionnaire in the Ohio study; and the almost competitive ratings by both girls and boys in Ohio of their own and the other sex's competence in using the electrical equipment. These differences could be a reflection of different educational environments (Stevenson & Stigler, 1992), but they also could be an indication that gender differences are decreasing (see Linn & Hyde, 1989), and that girls and boys are more aware of the issue of gender in participation and achievement in physical science and that the girls, perhaps, were more likely to challenge traditional stereotypes in this regard.

In both studies, the outcomes suggested that equity training in teachers' workshops, combined with activity-based lessons, affected students' attitudes and perceptions. Girls' enjoyment of and participation in physical science was raised to be equal to, or greater than, that of boys. Although on average girls still considered their own competence to be less than that of 'most girls', their confidence improved. The results were an encouraging step towards more equitable teaching of physical science in schools.

PART III: TOWARDS AN EXPLANATORY MODEL

The consistent findings from the two studies described above suggested to us that certain basic principles of school and classroom life which are important to equitable education in science (and possibly also mathematics) appear to transcend national differences. Our research demonstrated that, irrespective of culture, there are, first, similar antecedents for the beliefs, attitudes and perceptions held by both teachers and students about science and, second, similar factors interacting to determine the relationships between teacher behaviour and student behaviour in classrooms. Our results suggested that attempts to modify teachers' ideas and their level of knowledge and skills in science can influence their classroom behaviour, with corresponding changes in students' behaviour. Our findings revealed also that the involvement of students in hands-on, inquiry-based science activities (such as those concerning electricity in these studies), enables them to experience enjoyment and success, resulting in changes in their beliefs and self-perceptions about their interest and confidence in science.

We have considered the findings from the two studies in the context of an extensive review of related research (Kahle et al., 1993), which focused on three groups of variables relating to science and gender: teachers' beliefs, attitudes and perceptions; students' beliefs, attitudes and perceptions; and teachers' and students' behaviour in the classroom. We have proposed a model (Figure 1) which links these variables together, and which can provide a contextual framework for explaining gender differences in science-related attitudes, perceptions, classroom behaviour and learning outcomes. The model reflects the underlying premise that teachers' beliefs and attitudes are expressed through their behaviour, and hence there is no direct causal link implied from teachers' beliefs and attitudes to students' behaviour or students' outcomes. The model represents our attempt to explicate the interactions and cycles among variables which culminate in differences in observable outcomes of science, but our discussions with mathematics educators suggest that it also has considerable relevance for mathematics outcomes.

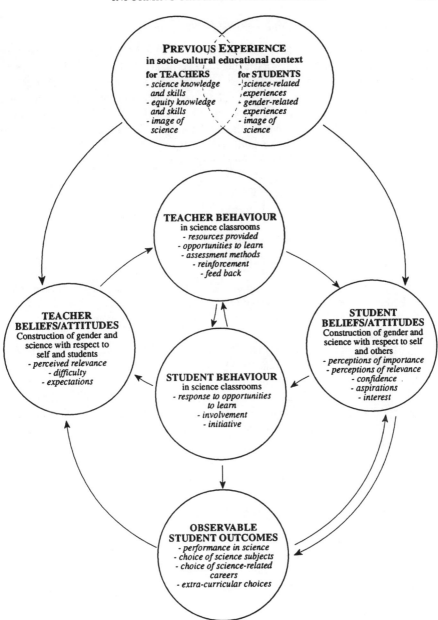

Figure 1.   A model for explaining gender differences in science-related attitudes, perceptions, class-room behaviour and learning outcomes.

Applications of the Model

We discuss here two particularly salient aspects of the model, with reference to earlier chapters in this volume. The first aspect concerns action to redress inequities in science/mathematics education. As is shown later in this chapter, the model suggests critical points for action or intervention, because such action can be conceptualised in terms of breaking some of the cycles such as, for example, those depicted in the model. The second aspect concerns the model's dynamic nature — it is intended to show that, for both teachers and students, the events and outcomes of yesterday's lesson (and other experiences) become part of today's previous experience, in a continuous process of adjustment.

The Importance of Outcomes of Science/Mathematics Education

In many senses, the observable outcomes of science/mathematics education play a critical role in the construction and legitimation of people's versions of reality. In large measure, it is these outcomes (observed, as depicted in the model, as gendered patterns of achievement, attitudes and participation) which have been responsible for the widespread concern about gender, science and mathematics that gave rise to the kinds of research and initiatives which are reported in this volume. For example, in Chapter 7, Keeves and Kotte document outcomes in achievement, attitudes and participation in science across countries involved in the first two International Association for the Evaluation of Educational Achievement (IEA) surveys in the early 1970s and mid-1980s. They interpret the variations to indicate that there are socio-cultural and educational variables which account for differences among countries and over time. In the context of the model, it can be seen that these patterns (as observable outcomes) become an important input to teachers' and students' 'previous experience'. Thus, there is a danger, as highlighted by the problematising of research by Johnson and Dunne in Chapter 5 and as alluded to by Leder in Chapter 8, that some research will contribute to the construction of the very 'gender differences' which it is attempting to address.

The Importance of the Socio-Cultural Context

At the top of the model, overlapping circles draw attention to the importance of the 'socio-cultural educational context' in framing teachers' and students' experiences. At the broadest level, these experiences occur in the same socio-cultural world. At a more detailed level, however, there are rather different subsets of that world, as Kahle and Damnjanovic (1994) show in their analyses by sex and race of the results from the Ohio study described earlier in this chapter. The experience of each person, whether teacher or student, is unique. It is his or hers to interpret and to use in the shaping and reshaping of beliefs, perceptions, attitudes and expectations, and ultimately in the attainment of the cognitive and affective outcomes of her/his involvement in school science.

Notwithstanding this degree of individuality, there are clusters of similar experience. As Jarvis points out in her description of young children's views in Chapter 3, from an early age, children's socio-cultural experiences influence the development of their images of science and scientists and, indeed, their ideas about who can become scientists. Furthermore, as shown by Haggerty in Chapter 2, these images endure. The preservice teachers in Haggerty's study struggled against their firmly-held perceptions, as it became increasingly clear that the science that they were experiencing was not consistent with their well developed, although not necessarily well articulated, views about science. In addition, as emphasised by Evans in Chapter 6, teachers' and students' experiences themselves are shaped by sometimes contradictory messages from the socio-cultural and educational context in which the individuals operate. Thus, as suggested also by Willis in Chapter 4 and by Kreinberg and Lewis in Chapter 15, there is at least one important precursor to action to make science and mathematics more gender-inclusive: teachers and students need to be helped to recognise and articulate the hidden and private processes of gendering which pervade the world of science and mathematics, scientists and mathematicians and science and mathematics educators.

The Heart of the Model – Teacher/Student Interaction

At the heart of the model lies the interaction between students and teachers in the classroom. Within this interaction, teachers' behaviour is conceived broadly to include the decisions which they make about major features of their teaching, such as the kinds of learning activities that they provide, the methods used to assess learning, and the nature of the feedback which they give to students. Many of these features have been explored in detail by the authors of chapters in this volume, including Murphy (with her insights into the gendered nature of assessment in Chapter 9) and Jorde and Lea (with their focus on curriculum materials and teaching strategies in Chapter 13). The model suggests that teachers' behaviour is shaped continually by two inputs: their own beliefs and attitudes about science/mathematics and about gender; and their interpretations of students' reactions to classroom happenings, which provide continuous feedback regarding the efficacy of teaching and learning activities.

Teachers' Beliefs and Attitudes. In relation to teachers' beliefs and attitudes, a number of chapters in this volume argue that the ways in which teachers construct the relationship between science, mathematics and gender depends on their knowledge and skills concerning science/mathematics and equity. However, as is demonstrated also by some earlier chapters, teachers' beliefs and attitudes are not fixed, but are cumulative and amenable to change. Change is difficult, as Haggerty's preservice teachers found but, through being given the opportunity to practise a different kind of science, these beginning teachers found that change can occur gradually. Alternatively, as reported earlier in this chapter, teacher change can be facilitated by exposure to credible communicators, in association with time for the teachers to reflect on the message conveyed by the communicators and to begin to 'own' that message. As argued by Koballa in Chapter 12, if teacher inservice workshops are aimed at changing teachers' attitudes or perceptions, attention needs to be paid to the credibility, attractiveness and power of the message giver. In relation to change, Jorde and Lea (in Chapter 13) also draw attention to the importance of considering teachers' needs. In their report of change in teaching patterns as the result of workshops associated with a refocussed, activities-based primary science curriculum, they emphasise that, even when change is mandated, teachers' classroom behaviour will change little if their needs are ignored.

Teachers' Interpretations of and Reactions to Student Behaviour. As shown in the model, teachers' attitudes, beliefs and behaviour can be affected by their interpretation of students' classroom behaviour and their observations of student-related outcomes. Thus, in the science/mathematics classroom, teachers provide learning activities and resources and interact with boys and girls according to how they perceive the students' needs. Teachers' interpretation of students' behaviour and their observation of students' outcomes, could provide feedback to confirm or challenge their beliefs and attitudes about gender and science/mathematics. Recall how the Ohio teachers in the study reported earlier in this chapter expressed surprise at how well girls performed the physical science activities. Clearly the girls' success was not consistent with the teachers' prior beliefs and expectations, and the teachers' classroom behaviour quite likely conveyed exactly those expectations to the students. However, partly as a consequence of student outcomes which were inconsistent with their expectations, the teachers were forced to adjust their beliefs. As indicated by Tobin (Chapter 10), however, teacher reactions are not always so constructive. He reports that some teachers try to dilute the effect of their (predominantly male) disruptive students in science classes by placing them in previously productive groups, thus disadvantaging other students, especially female students. Tobin's research underscores the importance of students' responsibility for supporting extant classroom practices. In this sense, he proposes a strategy for change, advocating that students be educated to reflect on their own learning opportunities and to challenge inequities in these opportunities.

Feedback to Students. The model also is conceived to be a dynamic one for students. Their classroom behaviour is depicted as an interaction between the teacher's behaviour (and the resulting learning environment in the classroom) and their own beliefs and attitudes in terms of how they construct the relationship between gender and science/mathematics. Thus, short-term outcomes of students' classroom behaviour, such as successfully completing an activity or passing or failing a test, not only give teachers feedback information to help them to judge the appropriateness of their own beliefs and expectations (as described in the previous section) but also give information to the students to make the same kind of judgements. An example of this dynamic interaction is found in Leder's descriptions of mathematics classrooms in Chapter 8, which reveal differences in student-teacher interactive behaviour which are remarkably consistent with differences in students' attitudes about mathematics. Her descriptions emphasise the importance of what happens in the classroom as a determinant of students' learning of and liking for mathematics. In a different context, Tobin's Chapter 10 demonstrates how the use of metaphors by a teacher can constrain the actions of a teacher and her/his students. In the case described by Tobin, a teacher's use of an 'entertainer' metaphor was associated with disadvantage to females, because the teacher interacted with his female students in ways which led to them feeling harassed and alienated from science.

Thus, as in the case of teachers, the ways in which students' beliefs and attitudes are developed are conceived to depend on their personal history and experiences and their interpretations of outcomes relating to science/mathematics and gender. Some successful action to make science/mathematics more gender-inclusive has addressed explicitly this link between students' attitudes/beliefs and their interpretations of observable outcomes. Farmer, for example, in the SOS program described in Chapter 14, focuses on a specific observable outcome and set of associated gender-stereotyped beliefs, namely, the lack of women in certain science and technology professions. Through her use of role models and other strategies, she challenged students' beliefs and provided them with an alternative view of what is possible for women in science and technology.

Conclusion: The Critical Role of the Teacher

Consideration of the model reveals the important role played by the teacher. The views which students develop from their out-of-school experience can be supported or challenged by the nature of the science/mathematics-related activities in which they participate at school. Change in students' views, their beliefs, performances and later further choice of science or mathematics can be affected by the way in which teachers choose to convert the curriculum into classroom activities. If change is to be attempted, teachers are positioned ideally to be change agents. Most of the initiatives described in this book are associated with assisting teachers to implement change in the ways in which science and gender are constructed. Further, as highlighted by Kreinberg and Lewis in Chapter 15, gender equity initiatives focused on teachers can take many forms, offering them support and practical help in changing teaching strategies to become more equitable.

As indicated at the outset of this volume, policy statements supporting more gender-inclusive mathematics and science education exist already in many parts of the world and at many levels. However, only by providing teachers with assistance of the kinds advocated by contibutors to this volume will the 'long shadow' between these policy statements and their implementation be shortened. Without appropriately targeted teacher inservice programs, the 'masculine strait-jacket' depicted by Harding earlier in this volume, will continue to constrain the development and progress of the more gender-inclusive science and mathematics education to which many now aspire.

ACKNOWLEDGEMENTS

Material in Parts I and II of this chapter is based upon research supported by the National Science Foundation under Grant No. MDR-88-50570 in the USA and by a grant from the Special Projects Program of the Commonwealth Schools Commission in the 'Education of Girls' section of Projects of National Significance in Australia. Any opinions, findings and conclusions, or recommendations expressed in this chapter are those of the authors and do not necessarily reflect the views of these sponsors.

[1]*Curtin University of Technology, Western Australia;*
[2]*Miami University, Ohio, USA*

REFERENCES

Bailey, B. (1992). 'The effect of teacher inservice on students' attitude and participation', Unpublished Doctoral Dissertation, Oxford, Ohio, Miami University.

Kahle, J.B., Anderson, A. & Damnjanovic, A. (1991). 'A comparison of elementary teacher attitudes and skills in teaching science in Australia and the United States', *Research in Science Education* (21), 208–216.

Kahle, J.B. & Damnjanovic, A. (1994). 'The effect of inquiry activities on elementary students' enjoyment, ease and confidence in doing science: An analysis by sex and race', *Journal of Women and Minorities in Science and Engineering* (1), 17–28.

Kahle, J.B., Parker, L.H., Rennie, L.J. & Riley, D. (1993). 'Gender differences in science education: Building a model', *Educational Psychologist* (28), 379–404.

Kahle, J.B. & Rennie, L.J. (1993). 'Ameliorating gender differences in attitudes about science: A cross-national study', *Journal of Science Education and Technology* (2), 321–333.

Kelly, A. (1978). *Girls and science*, IEA Monograph Studies No. 6, Stockholm, Almqvist & Wicksell.

Kelly, A (1981). *The missing half: Girls and science education*, Manchester, Manchester University Press.

Linn, M.C. & Hyde, J.S. (1989). 'Gender, mathematics and science', *Educational Researcher* 18(8), 17–19, 22–27.

McMillan, J.H. & May, M. (1979). 'A study of the factors influencing attitudes towards science of junior high school students', *Journal of Research in Science Teaching* (16), 217–222.

Ormerod, M.B. & Duckworth, D. (1975). *Pupils' attitudes to science: A review of research*, Slough, National Foundation for Educational Research.

Parker, L.H. & Rennie, L.J. (1985). 'Teacher inservice as an avenue to equality in science and technology education', *Contributions to the Third GASAT Conference*, London, Chelsea College, University of London, 226–234.

Parker, L.H. & Rennie, L.J. (1986). 'Sex-stereotyped attitudes about science: Can they be changed?', *European Journal of Science Education* (8), 173–183.

Rennie, L.J. & Parker, L.H. (1987). 'Detecting and accounting for gender differences in mixed-sex and single-sex groupings in science lessons', *Educational Review* (39), 65–73.

Rennie, L.J., Parker, L.H. & Hutchinson, P. (1985). *The effect of inservice training on teacher attitudes and primary school science classroom climate* (Report to the Commonwealth Schools Commission), Perth, The University of Western Australia.

Shrigley, R.L. (1983). 'Persuade, mandate and reward: A paradigm for changing the science attitudes and behaviours of teachers', *School Science and Mathematics* (83), 204–215.

Simpson, R.D. (1978). 'Relating student feelings to achievement in science', in M.B. Rowe (ed.), *What research says to the science teacher* (Vol. 1), Washington, DC, National Science Teachers' Association.

Stevenson, H.W. & Stigler, J.W. (1992). *The learning gap*, New York, Summit Books.

Vockell, E.L. & Lobonc, S. (1981). 'Sex role stereotyping by high school females in science, *Journal of Research in Science Teaching* (18), 209–219.

NOTES ON CONTRIBUTORS

MAIRÉAD DUNNE

Mairéad Dunne is a researcher in the School of Education, University of Birmingham, England. After teaching science in secondary schools in England, Mairéad worked in Kenya, where she established a Science Teachers' Centre and was involved in teacher education. She then lectured at the University of the South Pacific in Fiji where she was involved in curriculum development and consultancy in the countries of the Pacific Region. After a year at Curtin University in Western Australia, she returned to the United Kingdom to begin doctoral studies at the University of Birmingham, in which she examined mathematics teachers' and students' understanding of success in mathematics classrooms. Mairéad's research interests include gender and cultural issues in mathematics and science and she has published in this area.

TERRY EVANS

Terry Evans is Director of Research and Head of the Graduate School in the Faculty of Education at Deakin University, Australia. Dr Evans is an internationally known scholar in open and distance education and is the convenor of the biennial *Research in Distance Education* seminars hosted by Deakin University. He has co-edited six books on distance education including *Reforming Open and Distance Education* (London, Kogan Page, 1993) and *Critical Reflections on Distance Education* (London, Falmer Press, 1989) with Daryl Nation, and is the author of *Understanding Learners in Open and Distance Education* (London, Kogan Page, 1994). Dr Evans is on the editorial board of three journals: *Distance Education, Open Learning* and *Media, Technology and Human Resource Development.* He is an active researcher in open and distance education and is a recipient of Australian Research Council grants for his work.

BEV FARMER

Bev Farmer lectures in science and technology education at the Auckland College of Education in New Zealand. Her research interest is in biotechnology education and its classroom implementation. She believes that biotechnology education provides many opportunities for inclusivity and discussion of the many values issues that biotechnological solutions raise. One of Bev's projects has been the SOS (Skills and

Opportunities) program, which now continues with groups of educators running their own programs throughout New Zealand. The program has evolved into many forms but the focus continues to be the interaction of role models and students carrying out practical problem-solving exercises. The SOS team has produced a video *(Motorway Madness)* which demonstrates how this philosophy can be continued in the classroom. Video workpacks for both programs are available.

BARRY J. FRASER

Barry J. Fraser is Professor of Education, Director of the Science and Mathematics Education Centre and Director of the national Key Centre for School Science and Mathematics at Curtin University of Technology in Perth. Currently he is President of the 1500-member National Association for Research in Science Teaching in the USA and Executive Director of the International Academy of Education. He is author of *Classroom Environment, Windows into Science Classrooms, Educational Environments, Improving Science Education* and *Improving Teaching and Learning in Science and Mathematics*. At present, he is an Editor of the *International Journal of Educational Research* and the *International Handbook of Science Education* to be published by Kluwer.

SHARON M. HAGGERTY

Sharon M. Haggerty is Associate Professor of Curriculum Studies (Science) at the University of Western Ontario in London, Canada. She is an active member of the Gender and Science and Technology Association (GASAT) and was programme coordinator of GASAT 7 which was held in Ontario, Canada in 1993. Her research interests include gender and science issues and teacher education. Recent publications include papers on science teacher education and power relationships in science and science education. She is currently working (with Jazlin Ebenezer) on a science teacher education textbook.

JAN HARDING

Jan Harding graduated in Chemistry from the University of London. As her university studies were financed by a teachers' grant she trained as a secondary school teacher and taught in grammar schools for twelve years. After time off with her family, during which she organised and financed her own 'in-touch' programme, she returned to full-time professional work in teacher education. She worked part-time for Masters and PhD degrees in Science Education and finally identified as her research field the low participation of women and girls in science and technology. She has worked closely with UNESCO, the International Institute of Educational Planning (Paris), the Commonwealth Secretariat and with many different education systems and institutions in Australia, New Zealand, and Europe. In 1981 she was the co-founder of what has become the International GASAT Association (Gender and Science and Technology), of which she is the current Chair.

TINA JARVIS

Tina Jarvis is Head of the Primary Postgraduate Course for Initial Teacher Trainees at Leicester University, England. Tina taught in several inner-ring primary schools in Birmingham, England before becoming a lecturer in primary science and technology. She has particular responsibility for initial and inservice courses that lead to Certificates, Diplomas and MAs in Primary Science and Technology. She is author of *Teaching Design and Technology* (Routledge, 1993) and *Children and Primary Science* (Cassell, 1991) as well as articles on science and technology education. In 1991 she became the first woman to receive a Commonwealth Universities Development Fellowship to initiate research with centres for science and technology education in Australia, and she continued this work in 1994 with support from the British Council. Particular interests include the investigation of children's concepts of technology, science and scientists, and children's approaches to planning in a technological context.

JAYNE JOHNSTON

Jayne Johnston is the Learning Area Superintendent, Mathematics, in the Education Department of Western Australia. Prior to this appointment she was Consultant, Mathematics (K-12) with the Department, following her experience as a Head of Department of Mathematics and a secondary teacher. In 1989/90 she was seconded to the Science and Mathematics Education Centre at Curtin University in Western Australia, where she lectured in mathematics education and undertook research in gender and mathematics. Jayne was awarded a Commonwealth Relations Trust Fellowship in 1991 which took her to the Institute of Education, University of London, where she began her PhD studies. Her research project involves the development of a theoretical framework for a sociological analysis of mathematics classrooms. Her recent publications include papers in the areas of gender, constructivism, algebra and calculus.

DORIS JORDE

Doris Jorde received her PhD in Science Education from the University of California, Berkeley, in 1984. Since then she has been working in Science Education at the

University of Oslo, Norway, where she teaches graduate courses in science education. Through case study analysis, she has been studying primary science teaching with the aim of improving the current situation. Many of her projects have been international comparisons of educational systems. She has developed courses and curriculum for teachers at the primary level. Doris recently served on the committee which wrote the national curriculum for science for grades 1–10 in Norway which will go into effect in 1997.

JANE BUTLER KAHLE

Jane Butler Kahle, an international scholar in the area of gender equity, is the Condit Professor of Science Education at Miami University, Oxford, Ohio. Dr Kahle's research concerns the development and testing of a model that delineates the interactions among social, cultural, and educational influences on the career paths of women. As Co-Principal Investigator, she has made equity the cornerstone of Ohio's Statewide Systemic Initiative to improve the teaching and learning of science and mathematics. Recently, she has designed and implemented a study to assess the impact of the reform movement on student learning and on school change throughout the State of Ohio. She has written widely and has received national awards and several international fellowships for her research. She has applied her expertise as chairperson of the National Science Foundation's Committee on Equal Opportunities in Science and Engineering, as a member of the American Association for the Advancement of Science's Committee on Opportunities in Science, and as an expert witness for Congressional committees and task forces. In addition, she has served as chairperson of the Board of Directors of the Biological Sciences Curriculum Study and of the Gender and Science and Technology Association, and as president of the National Biology Teachers Association and the National Association for Research in Science Teaching.

JOHN KEEVES

John Keeves is a Professorial Fellow in the School of Education of Flinders University of South Australia. A major focus of Dr Keeves' academic attention has been the International Association for the Evaluation of Educational Achievement (IEA), and his contribution in this arena has been recognised by the award of Honorary Life Membership. He has worked in England, Sweden and Australia combining teaching and postdoctoral supervision commitments with his own research interests. His expertise led to consultancies in Brazil, Sweden and Australia and an ongoing international involvement in the theory and practice of education evaluation. Dr Keeves' recent publications include *Issues in Science Education: Science Competence in a Social and Ecological Context* (co-edited with T. Husén), Oxford, Pergamon, 1991; *Science Education and Curricula in Twenty-Three Countries* (co-edited with M.J. Rosier), Oxford, Pergamon, 1992; and, as Editor, *Changes in Science Education and Achievement: 1970 to 1984*, Oxford, Pergamon, 1992.

THOMAS R. KOBALLA

Thomas R. Koballa, Jr, PhD, is a Professor of Science Education at The University of Georgia. He regularly teaches elementary science methods courses and graduate level courses in science supervision and program evaluation. His research interests include science-related attitudes and persuasive communication. He has published widely in numerous scholarly journals and has been a presenter at many science and education conferences. Dr Koballa is Editor of *The Georgia Science Teacher Journal* and was recently elected President of the National Association for Research in Science Teaching.

DIETER KOTTE

Dieter Kotte, born in Hamburg, Germany, in 1956, was Assistant Professor at the School of Economics, University of Technology, Dresden and is now an International Consultant. He served in various functions to several IEA studies and bodies. He spent most of his studying and working years at the University of Hamburg, Twente University (Enschede, The Netherlands) and The Flinders University of South Australia (Adelaide). He is specialised in computer education and international educational comparative research.

NANCY KREINBERG

Nancy Kreinberg has been active in the movement to improve educational equity for all students since 1973, when she established the first "Math for Girls" class at the Lawrence Hall of Science, University of California at Berkeley. In 1975, she helped to found the Math/Science Network, which conducts yearly "Expanding Your Horizons" conferences throughout the US and abroad to introduce young women to careers in science and technology. In 1977, she created the EQUALS program and directed it for 20 students in mathematics. To date, over 75,000 educators in 36 states and 6 foreign countries have participated in EQUALS programs, which includes EQUALS workshops, FAMILY MATH, a program to help parents and children learn and enjoy mathematics together, 30 national sites that offer EQUALS and FAMILY MATH programs, and 15 curriculum books and a film. Ms Kreinberg's most recent book is *Teachers' Voices, Teachers' Wisdom: Seven Adventurous Teachers Think Aloud*.

ANNE LEA

Ever since Anne Lea started as a teacher, at 23 years old, she has worked with science, school and education. She has taught science from primary school to university level in Norway, and for some years gave a series of inservice courses in science for teachers. Anne has written several booklets in science which are used in Norwegian schools, and also made 10 TV programmes for the Norwegian Broadcasting System (NRK). Those programmes dealt with different science themes and were for primary school and lower secondary school. For some years Anne worked at the University of Oslo and did research and classroom studies in Norwegian schools. Currently, she is working at Oslo College, and her interests are mainly in teaching science to pre-school teachers and primary school teachers, and in research and work related to this field.

GILAH C. LEDER

Gilah C. Leder is a professor in the Graduate School of Education at La Trobe University in Melbourne, Australia. Earlier appointments have included teaching at the secondary level, at the Secondary Teachers College (now Melbourne University) and at Monash University. Her teaching and research interests embrace gender issues, factors which affect mathematics learning, exceptionality, and assessment in mathematics. She has published widely in each of these areas, including *Mathematics and Gender* (with Elizabeth Fennema, University of Queensland Press, 1993); *Assessment and Learning of Mathematics* (Australian Council for Educational Research, 1992); *Quantitative Methods in Education Research. A Case Study* (with Richard Gunstone, Deakin University Press, 1992) and *Educating Girls; Practice and Research* (with Shirley Sampson (Allen & Unwin, 1989). Gilah serves on various editorial boards and educational and scientific committees, including the executive of the International Commission on Mathematical Instruction and the Australian Mathematical Sciences Council. She is president of the Mathematics Research Group of Australasia. She is a frequent presenter at scientific and professional teaching meetings, and is currently working on several large research projects concerning gender and mathematics.

SUE LEWIS

Sue Lewis is the research and staff development coordinator in the National Centre for Women at Swinburne University of Technology in Melbourne, Australia, where she is developing and conducting gender inclusive programs for science and engineering staff in a number of Australian universities. Her research interests focus on understanding and diversifying the engineering and science learning and workplace environments. Sue has directed a number of gender and science research and intervention programs including the GAMAST Professional Development Manual *Gender Equity in Math-*

ematics and Science (co-authored with Anne Davies). Sue is a founding and active member of the educational organisation, the McClintock Collective, which has worked towards increasing the active participation of women and girls in science and technology education in Victoria since 1983. She is also active in the international Gender and Science and Technology (GASAT) conference and association.

PATRICIA MURPHY

Patricia Murphy is a Senior Lecturer at the Open University in Milton Keynes, England. She is responsible for producing distance learning materials for master's students and for the professional development of teachers in the areas of curriculum, learning and assessment and science education. Her research over the years has been focused in the areas of science and technology education with a particular emphasis on gender and assessment. She has published widely in this area. Her most recent book *A Fair Test? Assessment, Achievement and Equity* (1994, Open University Press) was co-authored with Professor Caroline Gipps.

LESLEY PARKER

Lesley Parker is an Associate Professor in the area of Academic Staff Development at Curtin University of Technology in Western Australia. She is also the Assistant Director of Australia's national Key Centre for Research and Teaching in School Science and Mathematics at Curtin. Her career in education has included several years as a teacher of secondary level science and mathematics, as a teacher educator and as an educational administrator in senior positions with the Secondary Education Authority in Western Australia. Her research focuses on structural curriculum change, professional development of educators and policy and practice in the area of gender equity. She has served on several major committees of enquiry into education at both State and National levels.

LÉONIE RENNIE

Léonie Rennie holds the position of Associate Professor at the Science and Mathematics Education Centre (SMEC) at Curtin University of Technology in Perth, Australia. She has a background in science teaching in Western Australian schools, and was involved in teacher education programs at The University of Western Australia before taking up a position at SMEC in 1988. Léonie is interested in the ways in which students learn science and technology, in both formal and informal settings, their attitudes about science and technology, and gender-inclusive assessment of cognitive and affective learning. She has published widely in science and technology education, and serves on the Editorial Board of the *Journal of Research in Science Teaching* and *The Australian Science Teachers Journal*.

KENNETH TOBIN

Kenneth Tobin holds the position of Professor of Science Education at Florida State University. After 10 years of teaching science in Australian and English schools, and preparing curriculum resource materials for teachers of high school science, he became a science educator in universities in Australia and the United States. His research interests focus on the reform of science curricula, teacher learning and change, and science teacher education. A recent past president of the National Association for Research in Science Teaching, he has published widely in science education and presented at International and National research conferences. He is an Editor for the *International Handbook of Science Education* to be published by Kluwer.

SUE WILLIS

Sue Willis is Associate Professor in Mathematics Education at Murdoch University in Western Australia. She began her career as a secondary teacher of mathematics before moving into curriculum development work. She has a range of professional interests all of which are framed by a commitment to social justice in and through education. Over the past decade she has undertaken a range of research projects which focus on gender reforms in schools including a long term study of the reception and effects of equal opportunity programs for girls on teachers and students and an analysis of the construction of gender within the new vocational agenda for schools. Her extensive publications have centred on this area. She has also been extensively involved in national level policy and curriculum development in Australia relating to the school mathematics curriculum and to numeracy education for adolescents and adults.

Science & Technology Education Library

Series editor: Ken Tobin, *Florida State University, Tallahassee, Florida, USA*

Publications
1. W.-M. Roth: *Authentic School Science.* Knowing and Learning in Open-Inquiry Science Laboratories. 1995   ISBN 0-7923-3088-9; Pb: 0-7923-3307-1
2. L.H. Parker, L.J. Rennie and B.J. Fraser (eds.): *Gender, Science and Mathematics.* Shortening the Shadow. 1996   ISBN 0-7923-3535-X; Pb: 0-7923-3582-1

KLUWER ACADEMIC PUBLISHERS – DORDRECHT / BOSTON / LONDON